# Catology

## The Weird and Wonderful Science of Cats

Stefan Gates

Hardie Grant
QUADRILLE

**To Tom. We miss you.**

# Contents

# Introduction

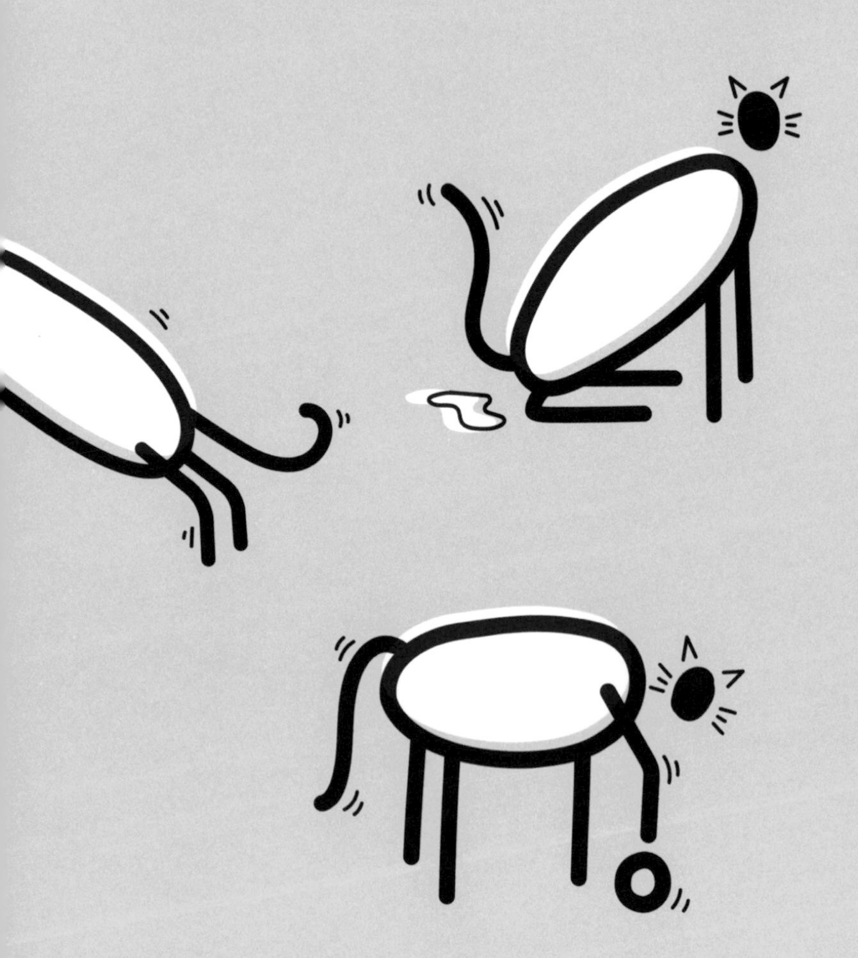

# A very unscientific introduction

C*atology* is a celebration of the 373 million* gorgeous, fickle, selfish, inscrutable, sofa-ruining, hairball-vomiting, bird-terrorizing furballs across the world that have inveigled their way into our lives and somehow made us fall in love with them.

Although this is a strictly scientific book designed to reveal the facts behind the domestic zoology project that is your cat, the relentless factmongery must occasionally pause to ground us in the real world where emotions interact with evidence, and I hope you'll allow me a brief moment now. I've owned two cats over the last 17 years. Tom Gates had a purr like a 1972 MGB Roadster, and was a big, food-loving pacifist hunk of love whom I treasured so, so dearly. Admittedly, he did once shack up with a feeder-neighbour for a few months, but he was gentle, loving and sensitive, and I cried a river when he died.

Our current cat Cheeky (not my choice of name, obvs), on the other hand, is a vicious, two-faced little monster who sleeps on my head, scratches me with open claws to demand stroking, licks my eyelids open every morning, and destroys all furniture with glee. Her favourite place to sit is on my keyboard (with sphincter placed square into my nose), and she has ruined many days of work on *Catology* by deleting files, rewriting my text and sending random emails (though her grasp of spelling is rudimentary at best). She loves a box, hates a vacuum cleaner, hisses like a snake when I try to squeeze the flea treatment on her neck, turns coy and needy after dark, and occasionally punches the dog. She loathes pretty much everyone, but saves a special level of hatred for our soppy hound Blue who, of course, loves her to pieces in return. But damn it, her very fickleness stirs in me that irrational, nebulous, ineffable, hormone-waterfall we call love.

I envy cats. I envy the instinctive, reactive simplicity of their lives and their contentment at sitting around for most of the day achieving bugger all. I envy their ability to suddenly switch from hedonistic pleasure-seeking to adrenaline-fuelled hunting, fighting or shagging. As far as we know, cats are unburdened by abstract thought, hopes, ambitions, guilt, indecision, self-doubt, moral and ethical confusion, and jealousy. Yet by just being in our houses, they give us a purpose, a pseudo-parental drive, a focus for our love and care, a drain for our money and a distraction from emotional, political, economic and romantic turmoil. And that's where the magic of owning a cat lies: the more we learn about them, the more we learn about ourselves.

In evolutionary terms cats have only recently wandered into our homes for warmth, shelter, food, and tickles behind the ears. In essence they remain wild predatory carnivores that have made

*Statistics on the global cat population vary wildly from 200 million to 600 million. Pet cat ownership is around 373 million, according to Statista, but that excludes a vast number of strays – probably twice as many again. There are thought to be 7.5 million pet cats in the UK, 94.2 million in the USA, and 3 million in Australia.

an extraordinary species-wide leap of faith to move in with us. And it's a great privilege to have them in our lives, because despite all the fickleness, aloofness, hissing, vom, trail of bloody mammals and ruined furniture, I suspect we still get a hell of a lot more out of them than they get out of us.

Thank you so much for reading this book. I'm a member of an odd but lovely little gang of people called science communicators, and we get a huge amount of pleasure not just from telling you amazing things, but from making learning exhilarating. You'll find us at science festivals, comedy clubs, in schools, on the telly, in pubs and in the kitchen at parties. If there's one thing we'd like you to take away from all this knowledge, it's that science can be fascinating, shocking, revelatory and often very, very funny. If you spot one of us on the street, do come and say hi. But beware: we are avid collectors of facts, and we've got so much to tell you.

## Note

There are dozens of cat species, including lions, leopards, cougars, ocelots and the gorgeous Pallas's cat, but we all know what sort of cat this book is really about, don't we? *Your* cat. For brevity, whenever the word cat is used, I'm talking about the domestic cat (*Felis catus*), unless I mention otherwise.

## Disclaimer

Nothing in this book is supposed to represent veterinary advice, behavioural advice or training advice. If you have any concerns about your cat, please visit a registered vet or animal behaviourist.

## Please ...

Be kind to animals and remember that their experience of the world and their perception is very different to ours.

# Chapter 02: What Is a Cat?

## 2.01 A brief history of the cat

**35–28 MYA**

Felidae (cat) family originates at end of Eocene/start of Oligocene epochs

**20–16 MYA**

*Pseudaelurus*, a prehistoric cat considered to be the first of the true cats, lives in Eurasia and migrates to North America

**8 MYA**

Distant cousins of domestic cats evolve in North America

**7–6 MYA**

The Felinae genera *Prionailurus*, *Otocolobus* and *Felis* (which includes today's domestic cat) diverge from a progenitor of the Felidae

**6 MYA**

The domestic cat's distant cousins migrate back to Asia

**2.5 MYA**

The impressively toothed *Smilodon* (a type of sabre-toothed cat) lives in North and South America before becoming extinct

**9500 BCE**

Agricultural societies flourish in the Middle East's Fertile Crescent. Agriculture = stored grain = rodents = cats for preying on rodents

**7500 BCE**

Date of burial site in Cyprus containing tamed cat remains

**5500 BCE**

Leopard cat (not a leopard but a small wild cat different from our common domestic species) is separately domesticated in China

**2890 BCE**

Ancient Egyptians worship Bastet, a lion-headed (and later cat-headed) goddess

**2000 BCE**

Cats kept as pets in Egypt

**450 BCE**

In Egypt, the penalty for killing a cat is death

**400–0 BCE**

While still revered, cats are now bred, killed and mummified on industrial scale in Egypt to be bought as religious offerings by temple visitors.

**300 BCE**

Date of British Iron Age hill forts containing cat and mouse bones – showing cats were introduced to Britain before the Roman conquest

**962**

Cat worship banned in Ypres, Belgium

**1233**

Pope Gregory IX links cats with Satan, causing millions of animals to be killed

**1658**

Cats continue to be demonized – cleric Edward Topsell writes 'the familiars of Witches do most ordinarily appear in the shape of Cats'

**1665**

Bubonic plague in London blamed on cats, despite being caused by disease-carrying fleas living on rats; 200,000 cats and 40,000 dogs are killed (thereby removing the rats' predators)

**1715**

The Age of Enlightenment begins, the church no longer dictates popular opinion as strongly, and cats become more popular as pets

**1817**

The last time live cats are thrown from the bell tower in Ypres, Belgium

**1823**

Pope Leo XII (1823–29) owns Micetto, a cat

## Cat Fertilizer

**In 1888 an Egyptian farmer discovered a mass grave of more than 300,000 mummified cats near two temples. Rather than being thrown away, they were stripped of their wrappings and shipped out to be used by farmers in England and the USA as a nutritious fertilizer.**

**1871**

First major cat show held at London's Crystal Palace

**1895**

First major cat show held at Madison Square Garden in New York

**1900**

Thousands of New York's feral cats are rounded up and gassed for 'humanitarian reasons'; children are paid one nickel per catch

**1910**

Florence Nightingale dies, having owned over 60 cats

**1947**

Cat litter becomes commercially available in the USA

**1975**

Cats banned from British naval ships

**2014**

Cat genome mapped

## 2.02 Is your cat basically a cute tiger?

Yes. No. Sort of.

Yes, your cat has an unnerving similarity to nature's most fearsome apex predator. Both tigers and domestic cats are solitary members of the Felidae family, they are obligate carnivores (see p142), fast ambush predators armed with vicious retractable claws and 30 teeth. They both share a love of climbing, scratching, grass-eating and rubbing against things to leave scent behind. In anatomical and physiological terms they share navigational whiskers, fur, a vomeronasal organ (see p86), a reflecting tapetum lucidum in their eyes (see p83), a vocabulary of spitting, hissing, snarling and growling (see p94), the ability to purr (see p95 – admittedly, tigers can only purr while breathing out), a habit of sinking their teeth into the back of their prey's neck to kill them, and, to cap it all, a stratospheric level of cuteness when babies. They both also sleep a lot, like catnip, enjoy playing in boxes, and share 95.6% of their DNA. So basically a tiger then?

Not entirely. While it's true that cats and tigers look very similar, you have to admit that tigers are a little bigger. **You could fit 39.625 average-sized cats into the average-sized tiger, and 77.5 cats into a particularly large male** – so it's no surprise domestic cats are a lower potential threat to prey. They are also much more vocal. And in evolutionary terms, the Pantherinae big cats split from the Felinae family of medium and smaller cats around 10.8 million years ago, so although related, they're not exactly brothers and sisters.

Of course, we all like to think of our cats as, well, big cats, so you'll be pleased to know there are more behavioural similarities than differences between Felinae and Pantherinae. A 2014 study published in the *Journal of Comparative Psychology* even concluded that **domestic cats share their three primary factors of personality with African lions: 'dominance, impulsiveness, and neuroticism'**. Touché.

## 1% Banana

**Cats share 95.6% of their DNA with tigers but that doesn't mean your cat is 95.6% tiger. We humans share 85% of our DNA with mice, 61% of it with fruit flies and 1% of it with bananas (not 50% as is sometimes reported), but that doesn't mean we are 85% mouse, 61% fly or 1% banana. Rather it means that all life on Earth evolved from a single cell 1.6 billion years ago and we all rely on oxygen, so we're all very distantly related.**

## 2.03 What's the difference between your cat, a wildcat and a feral cat?

Very little. Your cat (*Felis catus*) is basically an African wildcat (*Felis lybica*) whose ancestors have lived with humans over multiple generations – probably around 10,000 years. They are thought to have joined us soon after the development of agriculture when the grain that humans had begun storing attracted rodents. Wildcats were, in turn, drawn by those tasty rodents and bonded with nearby people, who may have offered scraps to keep them around and continue controlling pests. Subsequent generations of these cats became increasingly socialized, growing up alongside humans, where they could breed safely. **Domestic cats and African wildcats are so similar that it wasn't until 2003 that the powers that be finally ruled that the former were a unique wildcat subspecies called Felis catus**.

Wildcats look uncannily like large, striped tabbies with light sandy grey fur – because that's what they are. They are found across Africa, the Arabian Peninsula and the Middle East, most often in mountainous areas, but they also live in deserts such as the Sahara. There have been relatively few changes between wildcats and domestic cats since the two diverged, other than a slight reduction in size and an increase in some tameability aspects – specific wildcats were probably cherry-picked for their tolerance and affection towards humans and their environment.

Feral cats, on the other hand, are simply domestic cats that have run away or been abandoned to live in the wild. They're exactly as your own cat would be if she didn't return home every night.

Annoyingly for cat lovers, feral cats live very successfully without us, thank you very much. Once on their own, they tend to avoid human contact, don't take kindly to being touched, and return to preying on wildlife for their food, often with devastating ecological consequences. An Australian study found that **each Aussie feral cat kills an average 576 birds, mammals and reptiles per year, compared to a pet cat's average of 110**. Many people have tried to control feral cats with varying levels of success. TNR (Trap-Neuter-Return) programmes are thought to be the most humane method, but they consume vast amounts of resources, and their impact on the overall cat population seems to be minor. Feral cats are unusual in that, unlike solitary wildcats, they live in large social groups that allow them to share essentials such as food, water and shelter, and they even help to raise one another's kittens. Nevertheless, domestic cats, feral cats and African wildcats are remarkably similar and can all interbreed.

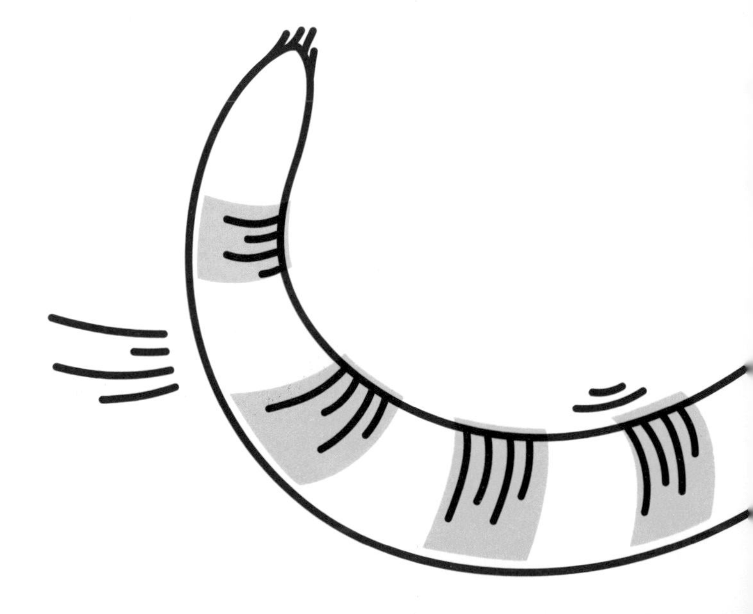

# Chapter 03: Catanatomy

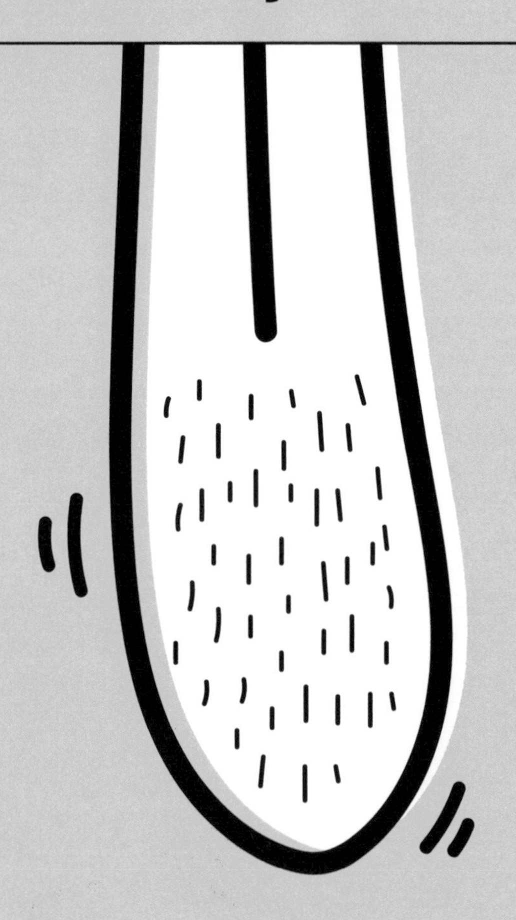

## 3.01 Why are cats' tongues so weird?

If you've ever been licked by a cat, you'll have noticed how bizarrely rough its tongue feels – like coarse sandpaper – and it doesn't take many licks before your skin is scraped raw. I know this because my cat licks my eyelids open at around 6am most mornings. Yes, she licks my eyelids open. And yes, it's as unpleasant as you can imagine.

A study published in the *Proceedings of the National Academy of Sciences* reveals that **cats' tongues are covered in hundreds of microscopic, rear-facing hooks called cavo papillae**. Researchers used CT scans to analyze these hooks and also filmed cats grooming themselves in slow motion to find out what's going on. It turns out the hooks are scoop-shaped and hollow, which allows each one to hold a store of saliva on the tongue that is then transferred on to the cat's fur as it grooms itself (presumably this is why cats feel, with some justification, that they never need a bath).

The researchers even examined lions' and tigers' tongues and discovered the same papillae structures, concluding that all cats' tongues work the same way. They write that 'the papillae wick saliva deep into recesses of the fur, and the flexible base of the papilla permits hairs to be easily removed from the tongue'.*

These tongue structures allow cats to clean themselves better than any other animal, despite their complex multi-fur structure (one of the many reasons cats usually smell better than dogs). Cats also use the extra surface tension created by the papillae when drinking

*The whole paper is free to access online and makes a fascinating read (https://www.pnas.org/content/115/49/12377).

– to pull water up into their mouths by lapping (in contrast, dogs use their tongues more like hammers to create a controllable splash). It's no excuse for torturing your owner at sunrise, but fascinating nonetheless.

## Mother of a Tongue

**If you have a spare few minutes, take a look at ragdoll_thorin on Instagram. Ignore his strangely blue eyes – Thorin has a very, very, very long tongue. There's no verifiable world record for the longest tongue on a cat, but Thorin's is a whopper.**

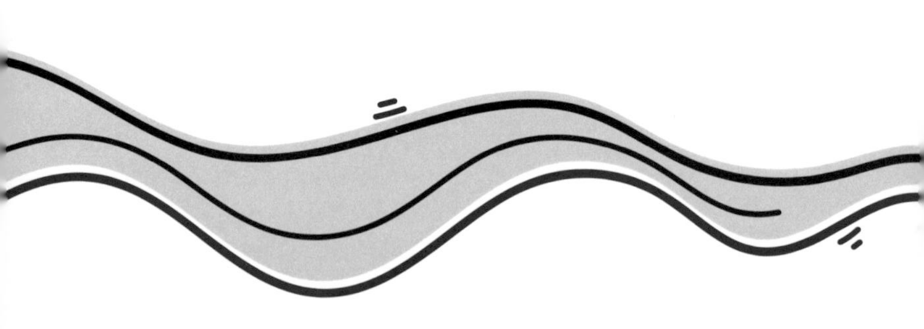

## 3.02 Why are cats so bendy?

Cats are agile hunting machines with around 20% more bones than humans packed into a lightweight and tough skeletal frame. Those extra bones (244 compared to our 206) are mainly found in the back and tail and help with speed, balance and agility. Cats' real speciality, though, is their flexibility, and this comes from the loose connections between their spinal bones and their floating front legs, which are attached to the shoulders by muscles and ligaments rather than sockets. This makes cats very bendy and helps with jumping, climbing, stretching, catching moving prey, and evading capture by larger animals. It also means that cats can twist themselves to squeeze through small spaces.

Because cats' collarbones aren't connected to any other bones, **they have very flexible necks and can turn their heads 180° each side for grooming**. If cats can fit their heads through a hole, many will try to follow with their entire body (search that on YouTube and you'll say goodbye to a few hours of your life). This bendiness is also a crucial component of their extraordinary self-righting skills (see p30).

### Long-jump Champ

**The longest officially recorded jump by a cat is 182.88cm (6ft), according to *Guinness World Records*. The cat's name was Alley.**

## 3.03 Cat sex

Right, sit down, you lot. I don't want to spend my morning talking about feline reproduction any more than you do but, funnily enough, Stretcher Fletcher has pulled another sickie, so you've got me. And no smirking at the back: no one's going anywhere until we've got through the whole sordid chapter.

Cats become sexually mature at around six to nine months and unneutered females usually come into heat each year between spring and late autumn. Their ovaries produce hormones in preparation, and they create scents and make calls that attract unneutered males, known as toms. **Females continuously seek attention (from humans as well as from cats) while on heat, rubbing at legs, on furniture, rolling on to their backs and making treading and stretching movements called lordosis**. They also assume the mating position – even when being stroked by humans – with front paws low and rear end up with tail in the air. And no, Yvonne, Mr Fletcher wasn't doing that in front of the headmistress at the Christmas disco.

Mating usually happens at night. The female gathers the local toms, who spray urine, fight between themselves and make their own yowling mating cries. The female is very much in charge of things, choosing the best suitor and angrily attacking any she doesn't take a fancy to. When she's made her choice, she allows the tom to mount her, whereupon he takes the scruff of her neck in his mouth – partly for stability and partly for safety so she can't bite him – and inserts his penis (which, incidentally, points backwards rather than forwards) briefly into her vagina before inserting some sperm. This is painful for the female as the **tom's penis has 120–150 rear-pointing hooks that scratch her vagina as he withdraws**, causing her to attack him immediately. Although it sounds horrific, the pain triggers the release of eggs from the ovaries and the female will quickly want to mate with the tom again, usually doing so several times, as well as with others. And no, Yvonne, it doesn't sound like an average lunchtime in the staffroom.

If the mating is unsuccessful and the female doesn't become pregnant, she'll go into heat again in a few weeks' time. If it does work, the pregnancy will last around 63 days, producing an average litter of three to five kittens (but it could be many more). The kittens may all be from one father, but the female could also have been fertilized by several toms, giving them correspondingly different markings. And yes, Yvonne, it's quite possible for one of Mrs Fletcher's kids to be ginger and still be Mr Fletcher's.

There, not too bad, was it? Are we all good? Great. Go on – get out. Not you, Yvonne. Laugh and the class laughs with you, but you stay on in detention alone. And don't hiss at me.

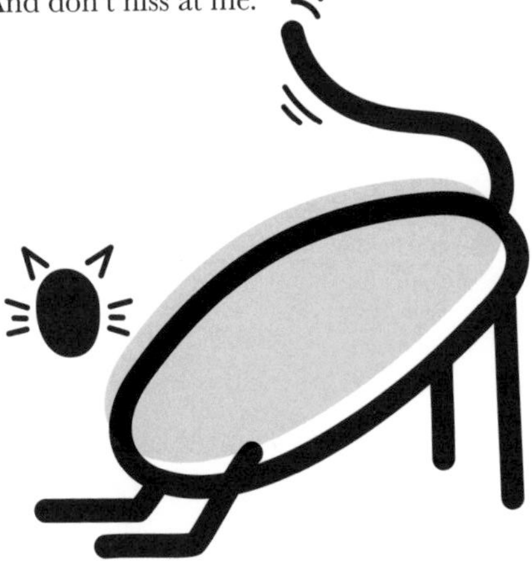

## 3.04 Can cats be left- or right-pawed?

Researchers at Queen's University Belfast found that cats do have lateralized motor bias paw preference (they're right- or left-pawed) when reaching for food, walking down steps or over objects. Overall, 73% had a preference for one paw or another when reaching for food and, while it was largely down to each individual cat's inclination, there was a clear preference among females to use their right paws and among males to use their left.

The team couldn't explain this female-male difference, but co-author of the research Dr Deborah Wells found a bizarre connection: the right hemisphere is more responsible for processing negative emotions and 'left-limbed animals, which rely more heavily on their right hemisphere for processing information, tend to show stronger fear responses, aggressive outbursts, and cope more poorly with stressful situations'. It's a bit of a stretch to conclude that left-limbed bias shows that male cats are more angry and neurotic than females, but there's definitely more to discover here.

Incidentally, lateralized motor bias isn't uncommon in the animal kingdom. **Ninety-five per cent of kangaroos are left-handed, 100% of pink cockatoos are left-footed, horses under the age of four tend to be right-nostrilled**, while cattle are left-eyed when looking at something unusual but right-eyed when looking at something familiar.

## 3.05 The science of paws and claws

Cats are digitigrade, which means they walk on their toes. It's a trait that lets them move quickly, quietly and with great precision – underneath all that fluff your cat is a hunter, after all. She also has a clever walking pattern called direct register, which means she puts her rear paw in almost exactly the same place that her front paw just left. This helps her to stalk prey more effectively by minimizing sound, and at the same time creating less of a track for other animals to follow. Among canines, only foxes have direct register.

**One of cats' many peculiarities is their walking gait, which they share with only two other animals: camels and giraffes**. It's a pacing gait, meaning they move both legs on one side before moving both legs on the other side. But as they speed up to a trot, they join most mammals in using a diagonal gait: their diagonally opposite rear and front legs move at the same time in a two-beat pattern. As cats accelerate into a gallop, they go through various asymmetric four-beat patterns – watch carefully and you can see your cat's legs all hitting the ground at different times.

### Paws

Most cats have five toes on their front paws and four on their back ones – a total of 18. The front paws have four digital pads with claws, plus one slightly redundant dewclaw further up their leg that doesn't touch the ground. They also have another pad called the carpal pad higher up the front leg, which is similar to our wrist pad. This can be used for extra traction or walking down slopes, but is thought to be an evolutionary cling-on that's no longer of much use. The back paws are simpler, with just four toes. There's also a large central metacarpal pad on the front and back paws, but as this doesn't have a claw, it's not really a toe.

The paw pads themselves are closely related to the hair on your head – they're covered in a rough, tough keratinized epithelium (human hair is made of keratin protein filaments, and epithelium is one of the four basic types of animal tissue, alongside muscle tissue, nervous tissue and connective tissue). The coarseness of cats' pads helps increase traction, making them less likely to slip. Underneath the pad is a thick grouping of fat and connective tissue (adipose and subcutaneous gelatinous tissue) that acts as a springy shock absorber to protect load-bearing limbs and ligaments.

## Hemingway's Six-toed Cats

**While a normal cat has 18 toes, polydactyl cats have a congenital abnormality that gives them extra toes – and it's not particularly rare. The writer Ernest Hemingway bred polydactyl cats on Key West in Florida after he was given a white six-toed cat by a ship's captain. Today, his old house on the island (now a museum) is still home to 40–50 cats with six toes or more on their front paws. Although their fat paws make them look as though they have thumbs, it doesn't seem to cause them any problems.**

## Claws

Claws are clever tools built for climbing, fighting, hunting and for scratching your best trousers. Almost all members of the Felidae family have them – lions' claws can grow to a terrifying 38mm (1.5in) in length. **Cats' claws are curled backwards (which makes them great for climbing up trees, though not so great for climbing down them again)** and, like paw pads, are made from keratin. Unlike human nails, they grow directly out of the phalange (finger) bones, and at their core they have an internal quick made of tissue containing blood vessels and nerves. You might occasionally see pieces of your cat's claws lying on the ground or stuck in your trousers, and this is because the outer layer of the claw, known as the sheath, naturally falls off every few months.

Your cat can retract her claws at will, and if she doesn't mind the sensation, you can gently press her metacarpal pad to watch them emerge. She controls her claws using ligaments and tendons, tensing her digital flexor tendon to extend them. When relaxed she keeps them sheathed almost entirely inside skin and fur to preserve their sharpness. They are curved for maximum grip, although they can over-curve if not kept sharp, which makes them more likely to get entangled in thick material. If she's getting overly long claws, you could carefully cut them back – but take advice from your vet and be careful in case you cut the nerve. If I tried it with my cat, she'd slice my nose off.

## 3.06 Why do cats have tails?

Cats' extraordinary tails are made of around 20 caudal vertebrae (depending on the breed) and they can control them to a remarkable degree. The vertebrae are linked by an intricate set of muscles and tendons that allows cats to move every part of their tail independently, right to the tip. Tails have a surprising range of communication uses (see p97), but are also handy for running, chasing, jumping and landing, as a counterbalance when making quick movements, and for walking on narrow surfaces such as the tops of fences. **Cats can also spin their tails in the air to help turn themselves upright when falling from a height** (see p30). You should never tug on a cat's tail: it's packed with nerves and is also important for controlling urination and defecation.

While tails are very useful, they aren't essential. Cats who lose their tails through injury invariably survive and adapt to live successfully. Strangely, Manx cats do not have tails but seem to retain agility (although breeding them is tricky, as having two copies of the tail-less gene also seems to result in most fetuses being spontaneously aborted).

**The Longest Domestic Cat Tail**

**Cygnus Regulus Powers, a silver Maine Coon cat, had a tail that was 44.66cm (17.58in) long, according to *Guinness World Records*.**

## 3.07 Why do cats have those evil-looking eyes?

One minute your cat's pupils are all big and round and come-on-gimme-a-stwoke and then the sun breaks through the clouds and they transmute into freaky slitty alligator eyes. There's only one word for it: vampiric.

While we humans have circular pupils that dilate and contract to regulate the amount of light that enters our eyes, cats have ingenious vertical slits that work like sliding doors. In the dark, cats' pupils slide open wide to let as much light in as possible and thereby look more rounded, but in bright light they close to a thin gap so as not to blind them. Cats share this anatomical quirk with snakes, geckos and alligators, and it allows their eyes to cope with a much wider range of light intensity than those of humans: **their pupils can undergo a 135–300-fold change in size between constricted and dilated shape** compared to humans' 15-fold change in size. This extra range means cats' eyes are better able to deal with the double challenge of night-time hunting and daytime self-preservation. A bit like vampires.

A 2015 study published in *Science Advances* analyzed 214 different species and discovered that eye structure is related to three things: the way animals forage, the time of day they are active, and their size. Grazing animals including horses, deer and goats have a similar ocular system to cats, but can rotate their entire eyeballs within the socket to keep the line of their pupils parallel to the ground to look out for predators. I know! This horizontal pupil system helps these animals cut out the sun from above so they can concentrate on the ground. But cats are generally predators rather than prey, and their style of ambush hunting often involves climbing, meaning vertical, rather than horizontal, contour range is more important for them.

## 3.08 Why do cats always land on their feet?

In 1894 French physiologist Étienne-Jules Marey made the first ever cat video using a specially made chronophotographic gun camera (an ancestor of the movie camera that looked like a hacked Bugsy Malone-style machine gun). He wanted to understand how cats always land on their feet, and his movie captured the answer: their amazing and beautiful self-righting mechanism. You can watch Marey's film on YouTube: the cat is held upside down and dropped from a height (well, it was the 1890s) and you can just about make out what's going on.

Cats' sense of balance relies on an extraordinary combination of vision, proprioception (a sense of body position and movement obtained from sensors in muscles, tendons and joints), and their vestibular sensory system (the balance and spatial perception mechanism in the inner ear). **Within one-tenth of a second of a fall, the cat's vestibular system analyzes which way is up**, and the cat turns its head to the ground, using its vision to judge where it's going and how far away the ground is. From here, it's all about biomechanics: the cat tucks its front legs in while extending its back legs, making it easier to spin its front to face the ground. (The cat uses inertia to control the spin in the same way that figure skaters do – by holding its legs and arms close or spreading them out.)

Then, with its front legs facing the correct way, the cat now switches to reposition its back legs, extending its front legs and tucking in its back legs so they can also twist to face the ground. This manoeuvre is often aided by a counter-rotating tail spin. All this is possible thanks to cats' extraordinarily flexible 30-vertebrae spines, which you can only really appreciate when you watch the spin in slow-motion. In higher falls, cats spread all four legs to increase wind resistance, parachute-style, which slows their terminal velocity to about 85km/h (53mph).

As the cat prepares to land, it extends its legs towards the ground and arches its back. On contact, the feet touch down and the back relaxes its arch, absorbing some of the impact and protecting the legs. Beautiful, no?

Cats need time to perform this self-righting manoeuvre so, counterintuitively the shorter the fall, the more likely they are to injure themselves. One 1987 study found that 90% of cats survived after falling from multi-storey buildings and only 37% needed emergency medical care. Cats that fell from seven to 32 storeys (32!) suffered fewer injuries than those that fell from two to six storeys. Amazingly, one cat that fell 32 storeys only suffered a chipped tooth and a small lung puncture and was on its way within 48 hours.

### Survivor Cat

**Cats can be highly resilient: 80 days after an earthquake struck Taiwan in 1999, a cat was discovered trapped in a collapsed building. It had lost around half its weight but it made a full recovery in a veterinary hospital, according to *Guinness World Records*.**

## 3.09 How many hairs are there on your cat?

The average cat has around 200 hairs per $mm^2$ (129,000 per sq in) and a skin surface area of approximately 0.252$m^2$ (2.71 sq ft) – for a 4kg (9lb) cat – giving it around 50.4 million hairs. It's a heck of a lot compared with us humans, who have around 90,000–150,000 hairs on our heads and about 5 million across our entire body. But it pales in comparison to other animals: **honeybees have 3 million hairs despite their tiny size, while beavers have 10 billion hairs. But even they can't compete with hairstreak butterflies and luna moths, which have 100 billion microscopic hairs each**.

Cats' fur is a wonderful, complex jungle of different hair types. In order of size from shortest to the longest they are: down hair, awn hair, guard hair and whiskers. The tiny down hairs create a soft, short insulating layer made even more effective by their microscopically fine waviness. Awn hairs are more bristly with thickened tips and make up the middle layer, protecting the down hairs and offering extra insulation. Guard hairs form a coarse outer layer to protect the underfur and keep it dry. Guard hair follicles also detect air movement and can control the hair itself to stand erect in response to anger and fear (piloerection). Highly sensitive whiskers are found mostly on the muzzle, ears, jaw, forelegs and above the eyes, and are used to judge wind movement and close-up touch navigation in the dark. For every 100 down hairs there are around 30 awn hairs and two guard hairs, although this depends on the breed. There are usually 12 whiskers on each side of the muzzle and differing numbers on the rest of the body. Maine Coons don't have awn hairs and the Sphynx has a light covering of down and no whiskers.

Your cat's fur protects him from injury and the elements, but its main function is to keep his body temperature in the right range of 38.3–39.2°C (100.9–102.6°F) – 2°C higher than humans'. This is called thermoregulation and fur makes a fantastic insulating layer to keep him cosy, although it also makes it more difficult for him to cool down (see p47).

**Cats have compound follicles (meaning many hairs grow from each follicle), and they secrete an oily sebum to keep their coats glossy and healthy**. They also create scent-filled moisture – not for cooling down but for communicating with other cats. Hair is made of the remains of tough keratin protein cells that are highly insoluble. This makes them wonderfully durable, but almost impossible to digest. Hence the hairball (see p46).

## 3.10 Why don't cats bark?

Search YouTube for the phrase 'barking cat', and you'll come across a wildly popular video of a black cat crouching perilously on an open window, barking at something outside. When someone disturbs the cat, its bark turns into something more like a plaintive yowl. Whether or not this particular bark is a fake, reports of cats making bark-like cries are relatively common – as are those of weirdly meowing dogs. None of them are exactly like the real thing (cat 'barks' often sound more like painful coughs), but it's clear cats can get quite close.

**Cats and dogs have the same basic machinery for barks and meows: their larynx, trachea and diaphragm are similar**. Barks are made by expelling air more forcefully through the vocal cords than required for most cat sounds, but some cats make them anyway – possibly because they're imitating a neighbouring pet, possibly because they're ill or confused, or possibly just to freak out the local canines.

So, if cats can bark, why don't they? Well, because they don't want to. Cats are solitary hunters that – except when mating – actively avoid meeting or communicating with other cats, so they don't like making noises that would give away their presence (the screeching a cat makes during a stand-off is a verbal tool for avoiding a physical fight, rather than the taunting shout of a boxer at a weigh-in). On the other hand, dogs are descended from social pack animals that benefit from lots of communication, of which barking is a particularly loud example. The trouble is that we don't know what barking means or what significance it has and it's quite possible that dogs don't know either.

## 3.11 Why do cats sleep so much?

Cats are lazy buggers, sleeping for up to 16 hours a day. By the time they are 10 years old, they'll only have been awake for three of those years. Weirdly, their brain continues to register smells and sounds for about 70% of that sleeping time in case they need to react to danger or a hunting opportunity. Even when they are awake, house cats are really very lazy indeed. **On average they spend 3% of their time standing, 3% of their time walking, and only 0.2% highly active**.

Cats sleep so much because they don't need to do a lot else. Their owners supply all their essential needs, so unless they feel like having a stroke or a poo, what's the point in getting up? Of course, they also have their hunting instinct to satisfy, and although cats were once thought to hunt at dawn and dusk, recent studies suggest there's a huge amount of variation between individual cats' schedules. While many cats are active at sunup and sundown, there are also lots of nocturnal hunters, as well as many that barely hunt at all. Nonetheless, as all cats are basically African wildcats that have been brought into our houses, they are ancestrally hard-wired to tend towards nocturnal hunting, so they are usually more active at night, and hence sleep more in the day to preserve energy for later.

Cats, like human teenagers, tend to expend energy only if absolutely necessary. If you're feeding your cat lots of lovely food, she'll adapt her sleeping patterns to your feeding schedule, but she's also less likely to need to be awake. Even then, her instinctive desire to hunt will still drive her out to chase prey every now and then (with varying levels of success), whether she requires the extra food or not.

## 3.12 How old is your cat?

Compared to humans, cats develop extraordinarily fast for the first two years of their lives, and then their development slows.

| Cat age | Human equivalent |
| --- | --- |
| 3 months | 4 years |
| 6 months | 10 years |
| 12 months | 15 years |
| 2 years | 24 years |
| 6 years | 40 years |
| 11 years | 60 years |
| 16 years | 80 years |
| 21 years | 100 years |

But how do you go about comparing the ageing of such wildly different species as humans and cats? Researchers look at developmental markers we both share, such as weaning, independence and sexual maturity, as well as behavioural changes.

Kittens shift from playing mostly with each other to also playing with toys and objects at the age of about 12 weeks, and this can eventually graduate to aggression between cats. They reach sexual maturity at about six months but it can be as early as four months. Depending on gender, young cats leave the family unit at one to two years, when they will often also start urine marking and other behaviours. As they age, adult cats play less and are prone to weight gain (much like me), while senior and geriatric cats undergo more behavioural, health and vocalization changes and are susceptible to a surprisingly similar set of health issues to humans.

In the wild, cats have a life expectancy of two to 16 years. **Domestic cats who stay indoors live an average of 13–17 years, while outdoor cats are thought to live two to three years less on account of suffering the dangers of cat fights and traffic accidents.**

## Cat Beats

**Cats live fast. Their hearts beat at 140–220 beats per minute (bpm) compared to humans' 60–100bpm. In contrast, pygmy shrews' hearts beat at 1,511bpm.**

# Chapter 04: Revolting Catanatomy

## 4.01 Why does cat poo smell so bad?

Let's roll our sleeves up and delve briefly into our cats' guts. On the face of it, cats have many of the same digestive mechanisms that we do: mouth, stomach, pylorus, duodenum, small intestine, gallbladder, pancreas, enzymes, liver, kidneys, colon, bacteria, rectum. But cats' digestive systems are much, much shorter, and specifically adapted to metabolizing meat, which is relatively quick and easy to break down (the transit time from food to poo in cats is only about 20 hours, compared to 50 hours in humans). Interestingly, cats have no appendix, an organ that was long thought to be a pointless evolutionary hangover in humans, but has recently been shown to protect beneficial bacteria in the gut.

Why does a meat-rich diet give rise to bad smells? Well, the breakdown of protein in the gut creates many smelly sulphurous chemicals. These include hydrogen sulphide and the putrid sulphur-containing methanethiol – the main culprits in eggy farts. Body-builders who consume lots of protein supplements are notorious for having very smelly farts.

But cat poo also has a secret ingredient that puts it in a different league. **A group of Japanese researchers have studied cat poo smell and concluded that it's due to an organic sulphur compound similar to one in white wine**. This cat-specific chemical is 3-mercapto-3-methyl-1-butanol (MMB), produced by the breakdown of an unusual amino acid called felinine that is produced by cats (but not dogs). MMB is a particularly putrid thiol (notoriously fetid, sulphur-containing compounds), and there's usually more of it in a male cat's poo than a female's.

Cats seem to know how pungent their poo is, which is why they often try to bury it. This isn't just out of innate tidiness or embarrassment – submissive cats may do it to avoid ensuring the interest of local dominant cats. Your cat may also bury his poo in your garden in deference to you as the dominant predator in the house, or simply because, as a kitten, he observed his mother doing it to avoid drawing attention to her vulnerable brood. Cats have occasionally been known to poo on their owners' clothes, perhaps as a symptom of separation anxiety.

And what about cat pee? Why does that smell so bad? Cats' antisocial nature is at odds with their evolutionary requirement to get together to mate. That's why scent is so important to them: it's a way for them to date without having to meet. Their urine scent marks communicate a huge amount about their physical condition, strength and healthiness, their readiness for mating, and even any familial links between them (an evolutionary turn-off as inbreeding causes genetic problems).

Fresh cat pee doesn't have the nicest smell in the world, but it has nothing on old tomcat pee, which invariably ripens to a pungent, putrid, ammonial honk. Once again, **felinine is mostly to blame – males produce around five times more of it than females, and the more high-quality protein a tom eats, the more there is in his urine**. This in turn marks him out as a better hunter and thus a more attractive mate for a female who wants her kittens to inherit the best genes. Cat urine also contains an unusual amino acid called isovalthene, and as both chemicals degrade by oxidation and microbial breakdown, they create secondary flavour compounds such as MMB, alongside disulphides and trisulphides, for that extra-special, fruity tomcat essence.

## Cat Litter

**Cat litter first became commercially available in the USA in 1947, and cat ownership increased as a result. It's usually made from bentonite clay, which clumps together when wet, effectively encapsulating the poo in a coat of clay and making it easy to dig out. A lot of this isn't biodegradable, though, and ends up in landfill, adding to the ecological burden associated with keeping pets. Biodegradable litter is made from wood pellets and various vegetable sources. Newspaper is an old favourite, but can end up pretty revolting to handle.**

## 4.02 Why don't cats fart (but dogs do)?

This really is my world*. **While humans can easily fart 1.5 litres (2½ pints) of gas a day, most cats never fart at all (despite having intensely smelly poo).** It all comes down to the cat's protein-exclusive diet and matching physiology. Most human fart gases are a by-product of the bacterial breakdown of vegetables combined with a small amount of swallowed air, and a tiny amount of pungent protein-derived volatile chemicals, which supply the merry guff's delicious flavours.

Simply put, a fart is made up of two things: a huge volume of gas derived from the bacterial breakdown of vegetable fibre, plus tiny volumes of powerful smell volatiles derived from the breakdown of protein. And there lies the secret: cats are obligate carnivores (see p142) – they eat loads of smell-producing meat but very few gas-producing vegetables. They do have the mechanisms for farting – a colon to ferment food and a nice, tight, circular sphincter through which to guff – but their relatively short digestive tracts are more designed for the easy breakdown of proteins in the small intestine than the lengthy and complex gas-generating breakdown of vegetable matter in the colon. They're just not up to the job of brewing a trouser trumpet.

Cats' colons are still fascinating, though. They have evolved for two main purposes: absorbing water and electrolytes from food, and controlling the consistency of poo. They do, of course, have a microbiome (the world of microbes that lives in our gut), but although this is vital for gastrointestinal health and water absorption (most of cats' moisture intake comes from their food), it's not particularly big compared to that of humans, and cats don't rely on it for nutrients.

Dogs, on the other hand, are carnivores with an omnivorous twist. They can eat a small proportion of vegetable matter, which needs breaking down, and thus naturally produce gas. Although we may think dogs fart more than humans, it's more the case that dogs simply aren't embarrassed by farting as we are, so do it whenever they fancy. In contrast, we tend to hold it in until we go to the toilet or are in bed asleep (or just under the covers and struck by the urge for a good ol' Dutch oven).

**So cats may occasionally fart but it's unusual – many vets say they've never witnessed a feline guff.** If your cat farts, it's possibly due to swallowed air or a microbial imbalance from a gastrointestinal infection, worm infestation – or a diet containing too much vegetation or milk (the breakdown of lactose sugars in the colon can produce gas). If switching to a more carnivorous diet doesn't sort things out, it's probably a good idea to see your vet.

**Fartology: The Extraordinary Science Behind the Humble Fart* by Stefan Gates (Quadrille, 2018)

## 4.03 Why is it always my turn to clear up the cat sick?

Cats are enthusiastic pukers even when healthy, and there can be many different reasons. Often they've simply eaten too much, too fast and then decided to empty the tank, in which case food is regurgitated without having had much time to mix with digestive juices. Tell-tale signs are the familiar firm, lumpy, tubular sludge sitting atop your best rug. Luckily, it's easily removed: grab a damp cloth (and take a brief gag), and you're done.

More revolting – and worrying – is actual vomit, where the food has been mixed with acidic gastric juices that have started denaturing proteins in the food, making it thinner, sloppier, more acrid, and much more effective at penetrating expensive deep-pile carpets. Cats often have a tougher time retching to get this out and it may be caused by allergies or something irritating the stomach, such as grass, carpet thread or sharp items. Although you know it's going to ruin your morning, requiring a pan of warm water and possibly rubber gloves, it's better out than in. My cat voms once every 3–4 weeks, and it's invariably due to overeating (she seems to function best on lots of small meals but sometimes forgets) or, if it's summer, occasionally due to grass.

Even more worrying voms are caused by disease, bacterial infections, viruses and parasites such as worms, which can be accompanied by more frequent vomiting than normal. If your cat pukes more than once a week, or retches a lot without producing anything, call that vet.

**We operate a fair system in my house: whoever discovers the cat sick first clears it up. So why is it always me?**

Why does no one else ever put their toes into warm, lumpy vom in the morning? And even if I'm last into a room, none of my family has managed to spot it. How can this be? Do they not look at the floor?

I also happen to be an expert at the sport of Actually Catching Our Cat's Puke in a Bowl Straight Out of Her Mouth – a skill that seems to get little recognition from my family. When I explain the details to them, they tend to leave the room.

## 4.04 Hairball!

It's harrowing to watch your cat throw up a hairball, but there's little to worry about. These slimy, stinky, gastric gifts are usually delivered through a rippling series of horrific gags and retches. Strangely, though, it's all part of a cat's normal functions.

Strictly speaking, a hairball is a trichobezoar ('tricho' means hair and a 'bezoar' is a lump of matter trapped in the gastrointestinal system). It tends to take the form of a tightly packed cylinder of fur and gastric juices, but can sometimes include food or other swallowed matter. **Cats are prone to hairballs as a result of their constant fur-licking combined with the peculiar nature of their tongues**, which are covered in hundreds of microscopic hooked cavo papillae (see p19) that pull loose hairs out, especially when cats are moulting. These papillae have a flexible base to help stop them getting clogged with hairs, but many do get swallowed and end up in the stomach.

The balls form when hair strands get trapped in the sticky gastric mucosa that line the stomach, and can't be shifted via the usual transit system. Once a few strands get stuck, others join it and build up. Eventually this becomes irritating and triggers a vomiting action that makes the abdominal muscles contract, shooting the lump through the oesophagus. And there she blows! As it gets squeezed through the oesophagus, the compression creates the familiar cylindrical shape and Bob's your hairball. It's the price you pay for silky fur.

It is possible to find remedies, and even anti-hairball foods, but some vets think they're ineffective or even harmful. More importantly, look out for any vomiting, retching and gagging that doesn't produce a hairball, which could indicate a problematic blockage.

## 4.05 Do cats sweat?

Not really. Cats have few sweat glands – in fact they have more of them on their paws than on the rest of their body. They have a few on their chins, lips and around their anuses, too, but these are more for moisturizing mucous membranes – to stop them from drying and cracking. **Cats aren't really built for sweating: their fur would stop the sweat from evaporating, and all that oily moisture would soon turn it into a matted, sodden, stinking soup of dangerous bacteria**. Not nice.

Your cat's fur acts as a thermal regulator that works to keep her warm rather than cool. Cats are hunting animals designed to be most effective at dawn and dusk when their vision and hearing give them an advantage – and temperatures are coolest. That's why all they bother doing during the day is sleep.

So how does your cat control her body temperature if she can't use evaporative cooling like we do? Well, sleeping is a good start – less activity means less cellular respiration, and less energy use. Grooming is also useful: as she licks her fur she leaves a small amount of moisture behind and as it evaporates, it cools her down. And there are lots of simple practical behaviours she can use, too, such as lying on a cold or shady surface. If she's really hot, she can pant like a dog to cool down, but this is unusual. Keep an eye out for your cat in the heat of summer – if you see her pads leaving wet prints behind, it's time to step in to cool her down by finding somewhere shady. But do remember that a cat's body temperature range is 38.3–39.2°C (100.9–102.6°F) – higher than ours by 2° (see p33). Just because you're feeling hot, she might not be.

## 4.06 Fleas!

Fleas. I can't help but be impressed by the little buggers. Like most living beings (with the exception of politicians) they aren't inherently nasty – they're just another species trying to get on with their busy lives, raise the kids, eat a balanced diet and get a warm (furry) roof over their heads.

The cat flea, *Ctenocephalides felis*, is the most common flea species on earth and it thrives in warm humid conditions such as those found on domestic cats, domestic dogs and in your home. Adult fleas are reddish-brown, 1–2mm (0.04–0.08in) long, but relatively thin, like they've been trapped in the closing doors of a lift. Unless you have a microscope you're only likely to see little dark grains on your cat's fur made up of adults, larvae, pupae and eggs.

Fleas much prefer to live on cats and dogs than other animals, and female adults can only reproduce by feeding on blood, after which they produce 20–30 eggs per day (they can produce up to 8,000 eggs before they die). In one to two weeks **the eggs hatch into larvae and feed on organic debris – mainly crumbs of poo that adult fleas have excreted**. The larva eventually spins a cocoon and becomes a pupa for a week or more before emerging as an adult flea, starting to feed on its host's blood, and continuing the cycle.

But there's more to fleas than cute baby pics: fleas cause adult cats few problems unless the infestation is severe, which can lead to dehydration and anaemia. But they can carry diseases such as tapeworm and cat flea rickettsiosis and these can also infect humans. Fleas don't bite – they insert their proboscis into a cat's skin and suck the blood out – but they do regurgitate digestive juices on to the skin (sorry), which can cause horribly itchy allergic dermatitis.

Once fleas are in your life they can be damn hard to get rid of, although humans are very much second choice of host if there's a furry pet around. If you do have an infestation, treat all animals in your house with topical flea treatment regularly. And if it's more than a small smattering of fleas, vacuum everything and then vacuum everything again before disposing of the vacuum bag immediately. Wash everything, especially pets' bedding, at a high temperature and hope for the best. If that doesn't work, it's time to call pest control.

# Chapter 05: The Very Weird Science of Cat Behaviour

## 5.01 What happens in a cat fight?

Cats were not designed to be sociable, and really don't want to share their garden with that evil Marmalade from next door but one. But that doesn't mean they want to tear strips off each other all the time, either – in fact quite the opposite. Neutered males are relatively non-confrontational compared to spayed females, with unneutered males the most likely to fight, especially with each other. But even they are wary of being injured and will deploy all of their vocal and body-language tools to avoid a physical confrontation – attacking cats are as likely to get hurt as defending ones. You may hear a huge build-up (often in the middle of the night), but even if there is physical contact, it tends to be just tense paw-slapping – basically 'handbags at dawn' – and sometimes the conflict gets resolved by a brief chase.

The rare occasions that do escalate into full-scale fighting usually follow a common structure. First, there's lots of posturing – back hunched, body turned slightly to the side, fur standing on end. Then the dominant cat slowly approaches the moaning submissive cat, turning its head on to the side as it advances, stalking close and wailing. There are sometimes long tense periods where the cats sit motionless, moaning, spitting, growling and howling. This is often the point at which the submissive cat walks away extremely slowly. But if the stand-off fails (or the cats feel equally dominant), one cat will start the physical fight by attempting to bite its opponent on the back of the neck. **The defending cat will immediately roll on to its back and kick repeatedly at the attacker's belly with both its back legs, while also biting.**

This is the moment when injuries are most likely. The attacker's initial bite almost always fails to connect, making it vulnerable and just as likely to suffer. The two may roll around biting, kicking and screaming, but this tends to break up quickly and the stand-off resumes, until one cat either launches another attack or backs down. The defeated cat signals submission by crouching low with its ears flattened as it creeps off, while the victor turns away at right angles and performs a symbolic sniff of the ground before walking slowly away. It's all as depressing, tawdry and lacking in glory as a fight between humans, really.

## Legendary Cats

### Grumpy Cat

**This miserable-looking American cat (real name Tardar Sauce) first hit the web in September 2012 when her owner's brother posted a photo of her on Reddit. This became wildly popular and was followed by TV appearances, books, calendars, commercials for Honey Nut Cheerios, a YouTube game show sponsored by pet food brand Friskies, video games, and over 1,000 official merchandise items. In 2014 she even starred in her own movie, *Grumpy Cat's Worst Christmas Ever*. Tardar died from a urinary tract infection on 14 May 2019, aged seven.**

## 5.02 Does your cat love you?

Brace yourself, it's time for your psychotherapy. The standard attachment experiment is called Strange Situation: a mother brings her one-year-old baby into a room full of toys, then leaves. A stranger comes into the room and then leaves. Finally mum enters again. The baby's response is either: 1) secure attachment: the baby cries when mum leaves, but is happy again when she returns; 2) insecure anxious-ambivalent: baby cries when mum leaves and has trouble being soothed when she returns; 3) insecure avoidant: baby doesn't appear bothered when mum leaves the room, although their heart rate and blood pressure monitoring show they are highly stressed. Around 65% of babies are in the secure attachment group, and the theory goes that attachment issues are defined by parenting style, which can have a significant impact on sexual proclivity, levels of psychopathy and relationship disturbance as the baby grows up.

The influence of attachment theory has waned among psychologists, but we'll hold a torch for it here because the Strange Situation experiment was recently adapted for cats, and the results were fascinating. On the first run, the cats all freaked out and the test had to be abandoned. So the researchers tried again using kittens and a restructured study, and **a gratifying 64.3% of them showed a secure attachment to their owner. Cats LIKE US!** But is that just a kitten thing? Do cats develop their legendary aloofness afterwards? To find out, the researchers carried out the experiment again a year later with the same group, and the results were even higher, with 65.8% showing secure attachment. Love? Maybe. Incidentally, only 58% of dogs show secure attachment.

Other evidence for cat love is the fact that cats rub against our legs when they greet us. In the few instances where families of cats live together they tend to rub up the social pecking order but not down, so kittens rub against mothers, smaller cats rub against larger ones and females rub against males, but rarely vice versa. Hence, they may see us as family, and possibly as their superiors.

Researchers at Oregon State University found that two-thirds of cats tested were 'securely attached' to their owners and that while attachment behaviour was 'flexible, the majority of cats use humans as a source of comfort'. This is contentious though: in 2015, **researchers from the University of Lincoln found that cats didn't demonstrate an attachment to their owner**. Research can be frustrating sometimes.

It all depends how you define love. Cats cry out to be fed, to be let in, to be let out, and to be stroked, and if you see dependency, control and manipulation as a form of love, your cat probably loves you. I've had girlfriends like that and it didn't take long to realize that the creeping sense of doom that accompanied the manipulation was a clear sign that what we had wasn't love, whatever the quality of the snogging. Another study from Oregon found that the majority of cats preferred human interaction to food, though it was pretty close. Still it's not quite love, is it?

Cats lick us, seek out our attention with upright tails (see p97), show contentment with our company by sitting on our laps purring, and tend to come back to our houses after we let them out. You might say that this is just convenient for cats – they use these tricks to ensure a regular supply of food and warmth. But in *Cat Sense*, John Bradshaw writes that 'Cats' attachment to people cannot be merely utilitarian; it must have an emotional basis', pointing to a study that

found levels of stress hormones in cats were lower when they were handled by humans than when they were caged. He adds: 'It is logical to assume that they regard their owners as mother substitutes'.

We tend to do what our cats tell us to, (so perhaps they wield 'affectionate dominance' over us rather than love us, much like a rubbish girlfriend or boyfriend), but wouldn't it be nicer if they actively needed us? Well, the University of Lincoln study concluded that cats are 'typically autonomous … and not necessarily dependent on others to provide a sense of security and safety'. Professor Daniel Mills, who led the study, said: 'I think cats do emotionally bond with their owners, I just don't think that at present we have any convincing evidence that this is a form of psychological attachment in the normal psychological sense.' The Lincoln researchers' results were so strong that they titled the paper 'Domestic Cats Do Not Show Signs of Secure Attachment to Their Owners'. Yup, capitalized and everything. Ouch.

If you want your cat to love/like you more, a fascinating 2019 study showed that spending more time with them makes them more attached to you, which may seem obvious, except for the fact that it only works when the attention is initiated and ended by the cat. Again, cats are very much like the worst girlfriend or boyfriend you ever had. The other trick is the 'slow blink sequence' – **one study showed that slow blinking makes cats more attracted to you**. I use this all the time when my cat needs soothing and it really does seem to work.

## 5.03 Why do cats love catnip?

Lions, tigers, leopards and domestic cats are all intensely attracted to the leaves of catnip (*Nepeta cataria*), a herb from the mint family with small pink or white flowers. When cats smell it, they often nibble and lick the leaves before exhibiting behaviour similar to females in heat, rubbing and rolling over the leaves, purring, drooling, and even playing out phantom chases. **The cat is basically high.** The response lasts for about 10 minutes before smell fatigue sets in and the cat becomes immune to the catnip for the next half hour or so.

The attraction happens in around two-thirds of cats, and it's a reaction to nepetalactone, a volatile oil in catnip leaves that they detect via the olfactory epithelium inside their nasal cavity. The olfactory receptors project a response to the catnip to two areas of the cat's brain: the amygdala, which mediates emotional responses, and the hypothalamus, which, among many important functions, releases hormones and regulates emotions. The hypothalamus causes the cat's pituitary gland to create a sexual response, so the catnip works as an artificial cat pheromone (a behaviour-altering agent). If a cat consumes a lot of catnip, it may exhibit manic behaviour, anxiety, or sleepiness, but this is rare, and catnip is generally considered harmless.

Humans don't respond to catnip in the same way, but it has been used as a herbal tea, and alternative medicine practitioners claim it can treat migraines, insomnia, anorexia, arthritis and indigestion. Weirdly not all animals love nepetalactone – many insects absolutely hate it and it's a remarkably effective repellent against mosquitoes, cockroaches and flies.

## 5.04 Are cats capable of abstract thought?

Let's deep this one*. There's very little research on cat cognition for the simple reasons that: a) cats don't care what we think; and b) they are completely useless research subjects. But what the heck, let's give it a go.

Abstract thought is the ability to think in non-specific conceptual terms (love, justice and ethics are abstract concepts) rather than purely factual ones. There's a lot of debate about whether cats – or indeed any other animals – can think abstractly. Some animals demonstrate a surprising ability to solve problems. This is believed to be abstract thought because it shows they're able to think something through before doing it. Chimpanzees sharpen tools for use as spears and use abstract reasoning to get to food; bonobo monkeys use sticks to fish for termites; wrasse fish use rocks to crack scallop shells; and both parrots and rhesus monkeys have shown rudimentary counting ability.

And cats? In a Japanese experiment researchers showed 30 cats a series of boxes, some of which rattled when shaken, and some of which didn't. When the boxes were turned over, some were revealed to have contained an object and others were revealed not to. However, a few boxes ran counter to rational expectation: some that had rattled did not spill out objects, and others that hadn't rattled did. It was these boxes, the ones that didn't fit the rational expectation, that most interested the cats, suggesting they understood the connection between sound and object, and hence had a basic causal-logical understanding.

*Teenage for 'discuss something philosophical'.

So, cats understand the abstract concept of gravity?

Whoa there, cowboy. Animals can understand particulars (the building blocks of knowledge, such as 'a rattling box probably has an object in it') – in fact they can be remarkably clever at understanding particulars. But that doesn't mean they understand universals (the characteristics that particulars have in common), which are usually abstract concepts. Hence chimps sharpen tools, bonobos fish with sticks, and **cats can have expectations about the physical world (that a rattling box contains something that will fall out when the box is upturned)** – but this is not to say they understand Newton's laws.

You could say that a cat hunting and stalking its prey involves abstract planning and anticipation, but this could just be instinctive and reactive – a genetically pre-programmed urge to stalk, chase and eat, rather than a sequence of abstract thoughts ('I'm hungry so I'd better consume some calories. What contains calories in the right form for my digestive tract and is within my physical abilities to catch? Mice. How shall I go about it? Well, I'll stay very still near a mouse burrow until one walks past …').

Cats are also believed to dream (see p61), and it's tempting to think dreaming involves abstract thought, but even this is probably just the replaying of memories. More positively, Hungarian researcher Ádám Miklósi found that cats are able to follow a pointed human finger almost as well as dogs can, demonstrating that cats can understand what another animal is thinking (though, annoyingly, none of my cats have ever been able to follow a pointed finger). But again, this is more particular than abstract.

Indeed, a test by psychologist Britta Osthaus showed that **cats are in fact a bit rubbish at problem-solving**. She attached pieces of food to strings in different setups and found that cats could tug a single string to get at food, but failed to choose the correct string when there were either two crossed strings or two parallel strings with food only attached to one. And rather shamefully, they performed worse than dogs.

So that's a pretty firm 'no' to abstract thought in cats, but why? Well, a capacity for abstract thought is both an advantage and a disadvantage. It's nice for us humans to appreciate all that abstract art, ethics, religion, literature and philosophy, but the flip side is that we get all the rubbish stuff too: perception of evil, repression, guilt, existential angst, and the contemplation of mortality. Be careful what you (abstractly) wish for.

## 5.05 Do cats dream?

Scientists have been unable to prove conclusively that cats dream because they are remarkably unforthcoming on the subject when asked. But it seems highly likely. They have similar brain structures to us, show similar low-voltage, fast-wave EEG (electroencephalogram) brain activity, and they experience REM (Rapid Eye Movement) when asleep – at the same stage when humans typically dream.

In 1959 French neuroscientist Michel Jouvet disabled the mechanism in cats' brains that inhibits movement during REM, and **observed sleeping cats raising their heads, appearing to stalk prey, arching their backs and even getting into fights**, leading many academics to conclude that they do indeed dream.

Indications of dream activity have been observed in most mammals, and humans are only in the middle of the scale. Armadillos and opossums experience some of the strongest REM patterns, whereas dolphins' are remarkably weak. There's no scientific consensus as to why we dream, but there are many theories – including that it helps us to process emotions, rehearse social and threatening situations, and to strengthen new memories. And don't get me started on the significance of human dreams – it's a murky world of deep pseudo-psychology, wellbeing-fruitcakery and TV sofa idiocy out there. If you want to freak yourself out, have a go at reading Freud's largely debunked *The Interpretation of Dreams*. Dirty boy.

## 5.06 Can cats feel happiness?

You'd think this would be the easiest question to answer in the book: 'My cat purrs when I stroke him, therefore he is happy.' Annoyingly, it's not that simple (cats purr when they're injured, too), which is why biologists generally avoid discussions about cat emotions like the flippin' plague. The difficulty is that cats, unlike dogs, aren't very expressive. Wildcats tend to be solitary animals with little need to express or share feelings. However, **MRI scans show that cats' brains have the same areas that generate emotions as humans, meaning they at least have the right mental machinery to experience happiness**. The question is, do they?

But before we even get to answering that, we have to tackle the difference between feelings and emotions. In broad psychological terms (and bearing in mind that there's no precise scientific consensus on definitions and no clear list of feelings vs emotions), feelings are the conscious, subjective experience of emotions, whereas emotions themselves are reaction experiences – often physical, biological states activated by the senses via neurotransmitters and hormones released by the brain. So, in very simple terms, emotions come first, and can give rise to feelings. Cats show clear symptoms of feelings such as stress (urinating on walls and pooing in the bed), anxiety (urinary tract infections such as cystitis), fear (especially of attack), boredom, panic and surprise, and these are all emotional responses. But that doesn't mean cats experience feelings as we do. For instance, it's a stretch to define the fear of attack from a dominant cat as 'hatred' of that cat, or to say that his feeling of pleasure from being stroked is 'happiness'.

A feeling such as **happiness is more than the emotion of pleasure – it's an experience of subjective wellbeing**. It's the ability to be conscious of the effect that pleasure is having on you. How does all this relate to cat happiness? Well, solid evidence of pleasure among cats is hard to come by, but neuroscientist Paul J Zak found that cats experienced 12% higher levels of oxytocin (sometimes known as the 'love hormone') after 10 minutes of play with their owners. (Dogs' levels rose by 57.2%, but everyone knows they're more easily pleased.) Cats also release hormones such as epinephrine (adrenaline) and cortisol when stressed, and endorphins when excited. So yes, cats almost certainly feel pleasure and pain, but pleasure isn't necessarily the same as happiness, and pain isn't necessarily the same as sadness. As the biologist John Bradshaw puts it in *Cat Sense*: 'We are aware of our emotions to an extent that cats almost certainly are not.' So cats experience pleasure, but not necessarily happiness.

And what about guilt? If you shout at your cat for tearing up the sofa or eating that sea bass you've been marinating in lemon zest and fresh thyme all day, she may look as though she's walking away with a guilty downtrodden look, with ears flattened and a bit of a hunch, but the reality is she's probably reacting with fear to your tone of voice. There are lots of funny YouTube videos of dogs looking terribly guilty, but research shows that if you talk to dogs with an angry voice, they look guilty whether they've done something wrong or not.

## 5.07 Does your cat know when you're unhappy?

Dogs have a well-studied ability to follow and react to expressions of human emotion and are especially responsive if we cry. They evolved from pack animals that live in complex social groups so you might expect them to notice and respond to emotional signals. On the other hand cats (with the exception of feral cats) are much more solitary and have little need for social interaction so you might expect they wouldn't be able to grasp human feelings. But not so: a 2015 study published in *Animal Cognition* found that cats are 'modestly sensitive to emotion'.

It turns out **your cat responds differently to you depending on whether you're frowning or smiling**. Cats are more likely to perform 'positive' behaviours such as purring, rubbing or sitting on laps when their owners are smiling than when they're frowning.

However, they don't seem to respond differently to strangers making the same expressions, implying your cat probably learns his ability to read your face as part of his developing relationship with you. Of course, this could be simple classical conditioning: you are more likely to show your cat affection or give him treats when you're in a good mood (and therefore more likely to be smiling), so they may simply be responding to that.

None of this necessarily means your cat is empathizing with you when you cry, but when it comes to cats, I'll take any crumbs of affection they offer.

## *Legendary Cats*

### Chief Mouser to the Cabinet Office

There is evidence of the English government having a cat-in-residence dating back to the 1500s, but it wasn't until 2011 that the official title of Chief Mouser to the Cabinet Office was first awarded to Downing Street cat Larry, a brown-and-white rescue tabby from Battersea Dogs and Cats Home. His lack of mousing aptitude became famous, but in October 2013 he managed to catch four mice in two weeks. Although nervous of men, Larry apparently made an exception for Barack Obama, whom he rather liked.

## 5.08 Where does your cat go once he's out of the catflap?

If you thought your cat's life was packed with adventure, exploring far and wide, prepare to be disappointed. Most cats do precious little all day, spending 66% of their time sleeping, 3% of their time standing, 3% of their time walking, and only 0.2% highly active. And they cover a surprisingly small amount of territory. One Australian study found that urban cats had a 'home range' territory size of between 100m² (1,100 sq ft) and 6,400m² (69,000 sq ft), which may sound like a lot until you do the maths: 100m² is an area 10m by 10m (33ft by 33ft) and even 6,400m² is only 80m by 80m (263ft by 263ft). Not very big at all.

Once they leave the house, outdoor cats are likely to find a high place where they feel safe and then simply sit and watch their territory. They may be tempted by a little ambushing if they see a small mammal or bird they think they can catch, but GPS tracking usually reveals how little, rather than how much, cats do. Fighting with other cats is extremely rare – they go out of their way to avoid conflict, and submissive cats that have failed to establish their own territory may spend most of their time watching out of windows and spraying urine indoors for fear of encountering a dominant cat outside.

**In 2019 a team of researchers from the University of Derby attached cameras to cats and studied the footage. The first thing they found was that 25% of their subjects hated the cameras so much that they had to be removed from the study.** The first useful thing they discovered was that when cats are outside they are highly alert and actively scan their surroundings for long periods. They often meet other cats but don't actively socialize, and two individuals will often just sit a few metres from each other for up to half an hour. It's not exactly Growltiger's Last Stand*, is it?

*Growltiger is the 'bravo cat who lived on a barge' from TS Eliot's *Old Possum's Book of Practical Cats* (1939), who spent his days fighting and terrorizing the inhabitants of the banks of the River Thames.

## 5.09 What does your cat do at night?

It's a common misconception that domestic cats are nocturnal (active at night) like their wildcat cousins. A 2014 BBC *Horizon* programme tracked cats' movement and found that **urban cats are more likely to be diurnal (active during the day), while rural farm cats are more likely to be nocturnal, but many cats – both urban and rural – are crepuscular (active at dawn and dusk)**.

Crepuscular cats benefit from their excellent low-light vision, which gives them an advantage over small mammal prey such as mice and rats in the twilight. Mice have poor eyesight but good movement-detecting peripheral vision and often rely on their whiskers to navigate, making them perfect for ambushing or stalking at dawn and dusk.

Not all cats bother to hunt – researchers from the University of Georgia attached cameras to a group of cats and found that 44% of them hunted wildlife, mainly at night, catching an average of two pieces of prey over seven day. Many were also likely to put themselves in danger: the main risky activities were crossing roads (45%), encountering strange cats (25%), eating and drinking substances away from home (25%), exploring storm drain systems (20%), and entering crawl spaces in which they might become trapped (20%).

So, why have domestic cats changed so much from their African wildcat cousins' nocturnal patterns? Well, it may be that domestication affected their activity times, a theory supported by a small study from the University of Messina, Italy. It found that cats' activity is highly influenced by human presence and care, and thus might not be genetically hard-wired at all.

## Legendary Cats

### Tabby & Dixie

Abraham Lincoln's cats. President Lincoln is said to have claimed that Dixie was 'smarter than my whole cabinet'.

## Legendary Cats

### Nala

The most popular cat on Instagram with 4.3 million followers. Nala was rescued by her owner Varisiri Mathachittiphan at five months old, and she's very cute, with round blue eyes, and her own cat food brand.

## 5.10 Cat divorce

One day, our gorgeous cat Tom decided to leave home and hang out with someone else. He wasn't lost because we saw him perhaps one day in four, and he looked healthy and happy. We, though, were distraught: we loved Tom with a passion. As the weeks went on, I noticed he was getting fatter. I eventually trapped him and took him to the vet, who warned of obesity-related diabetes and put a collar on him saying 'Please don't feed me'. But he continued to expand and it became clear: he had a Feeder. We were desperately worried that he would become more ill. Finally, though, we got a breakthrough. As the leaves began to fall in autumn we were able to see into our neighbours' gardens, and my wife spotted Tom lying outside a window with a full bowl of food next to him. He had only moved a couple of hundred metres!

The neighbour was a sweet man who desperately needed company. He knew he shouldn't feed Tom, but found it hard to stop: he was troubled, and loved Tom very much. But our family loved him too, and we really wanted him back. Of course, Tom could visit anyone he wanted, but this lonely man needed to stop feeding him, and as Tom was very food-focused, this was likely to sever their relationship. This continued on and off for two painful years until the man moved away and Tom moved back in with us without a meow of apology and shrank down to normal size again.

**Cats are notoriously fickle creatures and it's relatively common for them to simply leave home and live with someone else**. Their departure often coincides with the arrival of a baby, another pet, or a change in the living environment, but sometimes it's down to human intervention. Colin Butcher, director of The Pet Detectives, a UK-based agency that specializes in recovering stolen or missing pets, reckons that about half of all cats have a second home. He also says that pet seduction (whereby cats are adopted by being fed and let in) is one thing, but it can lead to theft.

Colin is an ex-policeman and his MO is to drop in on the adopter and persuade them to return the cat. As he says: 'I can be very persuasive'. He believes most cases end satisfactorily, although he has also come across serial cat collectors who amass multiple local cats and keep them fed and even locked in. For anyone with an outdoor cat, it's a tricky moral conundrum: cats are wild animals and no one has ever told them that they belong to you. Legally, if your neighbour takes a shine to your cat, you have to show that they intend to permanently deprive you of possession to prove theft. But perhaps even more upsetting is the fact your cat simply prefers someone else.

## 5.11 Are cats really able to find their way home from miles away?

There are loads of news stories about cats travelling huge distances to return home. In 1985 a cat named Muddy jumped out of a van in Ohio and arrived back at her home in Pennsylvania 725km (450 miles) away three years later. In 1978 Howie took a year to walk 1,900km (1,200 miles) home across Australia and in 1981 Minosch spent 61 days travelling 2,369km (1,472 miles) home to Germany. But we should be wary of thinking that cats have an innate homing ability. After all, thousands of cats go missing every day and never return, but such stories rarely make the news. The few that do make it home Incredible Journey-style are likely to be the exception to the rule.

There's little scientific research on cats' homing abilities, and most of what we have is very old. In 1922 Professor Francis Herrick separated a mother cat from her kittens by distances that varied between 1.6km and 6.4km (1 and 4 miles) and she always found her way back to them (thank God – naughty professor!). In 1954 **German researchers found that cats placed in a maze with multiple exits tended to choose the exit closest to their home. But how or why they did so remains a mystery.** They may well use their senses of smell, sound and vision to track themselves back, but they may also just search continually until they find themselves home. There are also tantalizing hints that dogs have geomagnetic sensibility (they tend to prefer pooing with their bodies aligned on a north–south axis), but whether or not cats have the same ability isn't yet clear.

## The Van Cat

**The Van doesn't, as its name implies, live in a small commercial vehicle. Rather it's a domestic cat breed that seems to love water and has been observed swimming in Lake Van in Turkey. Weirdly, the Van cat from Turkey is a different breed from the Turkish Van cat.**

## 5.12 Why are cats scared of cucumbers?

Unless you've been living in a hole you must have seen the YouTube videos of cats jumping out of their skins after a cucumber is sneakily placed behind them. When they turn and spot it, they invariably leap terrified into the air, arching their backs and either running off or inspecting it warily. Why?

Studies have found that cats have a fully developed sense of object permanence, as well as short-term and long-term memory, so it's no surprise they might be shocked by objects suddenly appearing. But more importantly, you need to remember that your cat is little changed from her African wildcat ancestors, which would have evolved to avoid dangerous snakes. When a cucumber appears out of nowhere, it doesn't take the biggest leap to understand that your cat might see it as a live creature – and its shape is certainly snake-like. **So it's highly likely that when you sneak a cucumber on to the floor behind your cat, she will jump out of her skin thinking it's a snake**. Which is exactly why you shouldn't do it to her for the sake of a funny TikTok.

## 5.13 Why do cats hate water?

Many cats turn nasty at the threat of a bath (which is a shame – they look hilarious when soaking wet), but are nonetheless fascinated by dripping taps, puddles, or the bath you're lying in.

Cats' relationship with water is complicated, but not all of them hate it. They don't need bodies of water to get food as they rarely hunt aquatic prey (other than my cat Cheeky, Slayer of Goldfish). They also drink relatively little, getting most of their water from their food. So it's understandable that they have little need or desire to jump into baths, ponds or lakes. There are owners who say their cats (especially Angoras) love baths and even swimming, and the Van cat is sometimes found splashing around in the Van lake in Turkey – but Vans are exceptions to the rule, and they are thought to have more water-resistant coats than the average cat. If you've got the cash to splash, a small cat water fountain is a good idea as many cats are mesmerized by water, even if they don't enjoy immersing themselves in it (though a dripping hose is likely to give as much fascination for a much smaller outlay).

And what about bathing? The truth is that most **cats simply don't need a bath as they're experts at self-grooming**. They have evolved special hooked cavo papillae on their tongues to help with cleaning (see p19) and invest a fair amount of time in it. They've also worked pretty hard to get the smell of their fur just so, what with their endless rubbing against posts, secretion of scents, and all that competition-level fur-licking. Mess with that at your peril, human.

## 5.14 Why does your cat scratch the damn sofa?

Surprisingly, the answer is not to keep his claws sharp. There are several reasons for sofa scratching: on the simplest level cats are very tactile and love padding, kneading and stretching their claws, as well as depositing little patches of scent from their paws. More importantly, their claws are covered by protective keratin sheaths that are constantly regrowing and need to be shed every three months or so (see p27). Scratching helps them get rid of the old ones. These sheaths aren't quite the same as human fingernails, which grow constantly to protect the ends of our fingers, but they're not far off, and you may occasionally find a shedded one lying around.

**The most hilarious solution to sofa-scratching is claw caps – tiny fake sheaths you can get fitted to your cat's nails**. Depending on how bitter you're feeling about your shredded furniture, you can buy caps in all sorts of lurid fluorescent colours for maximum cat humiliation. You'll need to replace these every six weeks, but they may stop the scratching. A better solution is to buy or make a scratching post (preferably in stringy coir, or the same material your cat most enjoys destroying) and then place it by your knackered settee to displace the activity.

## 5.15 Why do cats get stuck up trees?

Cats love sitting in high places because it makes them feel safe. Even though they're successful predators, they're also prey for dogs, bigger cats and other large mammals. Sitting in a tree or atop the kitchen door where they can keep a beady eye on their territory and the stoopid family dog makes a lot of sense: there's no one else to disturb them, and cats really value their peace and quiet, especially from each other.

Trees are also a great place to chase birds, and this is where cats can run into problems. The excitement of the chase sometimes overrides any concern about the consequences of climbing too high …

The thing is, **cats' claws are curved backwards, so while they're great for climbing upwards, they're less useful for slowing a headfirst descent**. Cats are aware of this, but they aren't comfortable coming down backwards, so when a cat is stuck up high, it can get too distressed to come down. The reality is that cats are invariably able to scramble down on their own, even if their descent is pretty ungainly and scary to watch.

So what should you do? First, DON'T get a tall ladder. Climbing up high to handle a panicking cat is a great way to fall and hurt yourself very badly. It's generally thought that unless your cat's in real shock, he will eventually make his own way down given enough time and motivation (perhaps a rattle of his food bowl), and that you are likely to be more scared about the situation than he is. Much better to find yourself a rug, some cat food and a book, and settle in under the tree until he feels calm and bored enough to come down himself. Only if he stays stuck for many hours should you bother the fire brigade. And if you do, I'd advise that you bake a particularly fine cake ready for their arrival.

## 5.16 Why do cats love sitting in boxes?

Maru is a celebrity Scottish Fold cat from Japan. He loves a box. The video of him jumping into a box has had 10 million views. Maru jumps in the box. Maru jumps out of the box. Box falls over. Maru loves a box. Maru is cute. That's it, really. Ten million views. My cat will get into any box, or indeed anything that resembles a box such as a sink, bag, oven or washing machine. She'll even get in the mixing bowl I'm about to use for making bread.

Why boxes? Well, there's no evidence-based research to help us, so we have to rely on opinions, and here are some of the most convincing:

1. Cats are ambush predators and boxes give them a good hiding place from which to pounce on prey, while also concealing them from the stoopid family dog. The problem with this theory is that cats also get trapped in boxes, and cats hate being trapped, so they are taking a risk by getting in one. Presumably the hiding desire outweighs the fear of being trapped.

2. Cats are curious and boxes are places that need exploring.

3. One of the strongest theories is the feline equivalent of the 'if I put my head back under the duvet all my problems will go away' principle. A Dutch study in 2014 showed that **cats that were given a box to hide in after arriving at an animal shelter were much less stressed and got used to their surroundings – and humans – faster than cats not given a box**. Hiding in the box helped the

cats cope with their new environment. This fact squares well with cats' general inability to socialize: rather than tackling a changed environment, they simply prefer to avoid dealing with it – and the approach seems to tangibly improve their lives. Essentially, they like the box because they're the only ones in it.

## 5.17 Why do cats make great mums but terrible dads?

Cats are solitary, though domestic female kittens do sometimes continue to live with their mothers as long as there's enough food and no human intervention (males leave the family unit at around six months). When these remaining females have litters, they will usually, very sweetly, share caring responsibilities between themselves: a formidable gang of aunties.

Males rarely help with raising kittens. After all, females are likely to have mated with multiple males (which is why kittens from the same litter may have many different characteristics, see p23), so in evolutionary terms they can never be sure they're helping their own lineage. As they see it, they're better off trying to sire as many young as possible. Male cats have also been known to kill kittens that aren't related to them. This can cause the mother to come on heat again and thus allow the male to mate with her, prioritizing his own evolutionary line. For this reason, females are unlikely to want males anywhere near their families.

It's not quite as simple as 'mummy cat good, daddy cat bad', though. More shocking than male infanticide is **maternal infanticide, whereby a mother kills and eats one of her litter and then carries on caring for her other offspring as if nothing has happened**. This is not uncommon, and is thought most likely to occur when the mother senses a kitten is ill or deformed. In the wild, getting rid of the runt of the litter leaves more food and protection for the rest, improving their chances of thriving. Eating the kitten may seem beyond the pale, but why would a hungry and stressed mother cat pass up some decent extra nutrition?

The real tragedy here is that the mechanism that triggers a mother cat to think a kitten is weak or ill is very sensitive and can be activated by factors unrelated to the kitten: an uncommon smell near the kitten, unexpected behaviour, or even a vibration.

## Mighty Mammas

**Born in 1935, Dusty the tabby, from Bonham, USA, produced 420 kittens during her life. Her last litter was a single kitten, born on 12 June 1952. The largest litter ever recorded was 19 kittens, born to a Burmese-Siamese cross in Kingham, UK, according to *Guinness World Records*.**

# Chapter 06: Cat Senses

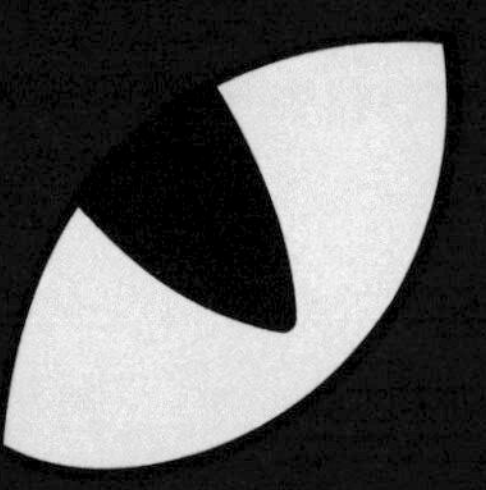

## 6.01 How do cats see in the dark?

Cats' eyes are beautifully adapted for hunting, especially in low light conditions. They are enormous in relation to the size of their heads – almost as large as human eyes – and their pupils can expand to three times the size of ours, allowing them to access much more of the available light (see p.29), perfect for a nocturnal killer.

But cats' secret ocular weapon is their **tapetum lucidum, a greenish retroreflective layer behind their retina that bounces light to the back of the eyeball, effectively allowing 40% more light into the eye**. This allows them to see in 0.125 lux (a measure of illuminance – by comparison humans can only see in 1 lux). Cats share this useful bit of kit with crocodiles, sharks, dogs, rats and horses, among others. If you point a torch at your cat's eyes in the dark, they will shine back green because some of the light is reflected by the tapetum lucidum and spills out from the retina. You can't see the tapetum lucidum in daylight as your cat closes its pupils into a narrow (and slightly creepy) slit.

But there's a downside to having eyes specifically engineered to hunting in low light – daylight vision suffers. Not only do cats have a much less detailed visual picture in daylight compared with humans, they are also nearsighted (they can't see faraway objects well) and longsighted (they can't focus on anything closer than 25cm/10in). Indeed their lens system is so cumbersome that cats often don't even bother to attempt close focus. Whereas 'normal' human visual acuity (clarity or sharpness) is 20/20, cats' is 20/100 – meaning they must be 20ft (6m) from something to see it as clearly as a human with normal eyesight does from 100ft (30m).

The photoreceptor cells that detect light at the back of cats' eyes are also different from those of humans. Although both cats and humans have rod and cone receptors (rods detect black and white intensity and cones detect colours), **cats' eyes have a much greater proportion of rods to cones compared with humans, which makes them very sensitive to light and dark but much less to colour**. Cats can sense blue and green but lack the cones to see red, and so have little interest in colour. Presumably they gain little evolutionary advantage from good colour vision as it isn't particularly useful for hunting. Instead their eyes have adapted to enhance the visual tools they most need for catching small animals.

Cats also have a higher flicker-fusion rate than we do, although it does depend on breed. This means that the visual cortex in their brains can compare images coming into it at around 100 frames per second – much faster than the 60 frames per second a typical human can process. As a result, cats can detect small movements better than us, but see old-style TVs and fluorescent lights with a flicker.

## *Legendary Cats*

### Lager Cat

**Karl Lagerfeld's cat, Choupette, is still famous after her owner died in 2019. She has a thriving Twitter account, an agent, and of course Lagerfeld may or may not have left some of his $200 million (£150 million) fortune to her.**

## Extra Eyelids

As well as excellent night vision, cats have a nictitating membrane, a translucent third eyelid that slides in from the side to clean and protect their eyeballs. Many birds have these – they're particularly prominent in turkeys – and they're also found in dogs, camels, aardvarks, sea lions (who only use them out of water), fish, crocodiles and other reptiles. My favourite nictitating membrane fact is that woodpeckers tighten theirs a millisecond before they whack a tree with their beaks to protect their retinas from shaking injuries. You'll struggle to see your cat's nictitating membrane in normal circumstances (if it is visible, she may be in poor health), but if you gently open her eye when she's asleep, you should be able to spot it. Good luck with that, though – my cat would probably tear my nose off.

## 6.02 How good is your cat's smell?

Cats don't have quite as good smell skills as dogs, but they still beat humans hands down. The mechanism is the same as ours: cats breathe in air and some of the flavour volatiles (smell-carrying molecules) in that air reach their olfactory epithelium – the area of the nose dedicated to sensing smells. This is five times bigger in cats than in humans and contains millions of odour-sensitive nerve endings covered by a thin layer of mucus. The flavour molecules dissolve in this thin layer and interact with some of the hundreds of different nerve endings to create a signal that is then sent to the brain. The different nerve endings detect different molecules (although the mechanism for this is poorly understood), and the brain uses this information to assess the overall smells.

Cats need their powerful sense of smell to track prey, but also to understand other cats' smells, usually left behind in urine, faeces or scent secreted from various glands. These give information about age, health, sexual status, but also mark territory to help cats avoid each other.

**Cats also have a separate second smell detection mechanism called a vomeronasal organ (VNO).** This is hidden in the roof of their mouths and accessed by two tiny tubes behind their upper incisors. Unlike the olfactory epithelium, it's a sac full of fluid and chemical receptors that senses the smell of molecules dissolved in saliva. Two sets of minute muscles pump the saliva in and out. Cats only use their VNOs occasionally – in social situations to detect the scents of other cats (which usually communicate sexual information). You can normally tell when they do because they have to pull a strange facial expression called the Flehmen response – it's a slight gaping sneer in which their lip pulls up to expose their top teeth, their mouth slightly opens, and their tongue hangs down. Horses and dogs have a similar action.

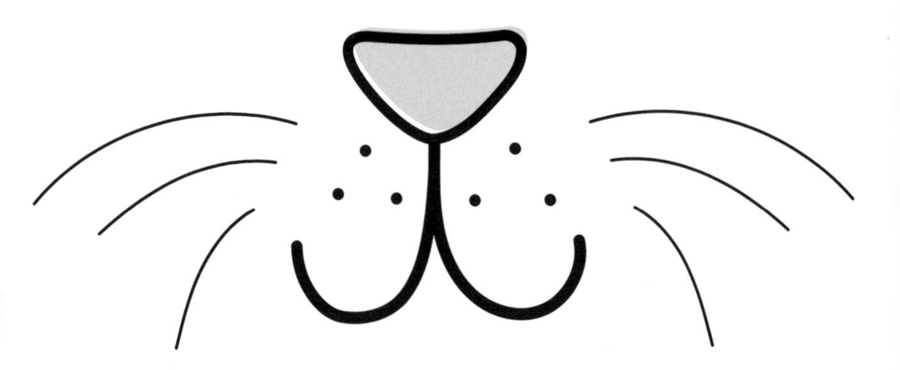

## 6.03 How good is your cat's sense of taste?

Cats have a relatively poor sense of taste, with only 470 taste buds compared to dogs' 1,700 and humans' 10,000. Their carnivorous diet means they don't need to be attracted to sweet-tasting fruits and vegetables. Instead their sensory system concentrates on the salty, bitter and sour tastes of meat on their tongues' papillae. Cats' inability to taste sweetness is hard-wired into them through a defect in one of the genes that codes for the T1R2 protein (part of the sweet taste receptor). This mutation happened early in cats' evolution, leaving them unable to feel a natural preference for glucose, sucrose and fructose. Metabolisms are energy intensive, so to preserve resources cats' bodies don't bother producing the sugar-digesting enzyme sucrase, because they don't eat sugars – or indeed any of the starchy vegetables that omnivores like us break down into simpler sugars. The downside is that if they do drink a sugary substance, they don't know it's sweet and, without the digestive tools to break it down, could become sick.

A 2016 study published by the Royal Society found that cats are more interested in the protein-to-fat ratio of foods, rather than their flavour. They can sense this ratio (although it isn't entirely clear how) and regulate their body's needs accordingly, favouring a balance of 70% protein to 30% fat. The researchers came to the extraordinary conclusion that in the long term, **nutritional balance was more important to cats than flavour, and that they actively eat a diet that matches their needs**. Which on one hand is a little sad, but on the other, would be really useful to those of us humans who are so distracted by flavour that we find it almost impossible to eat as healthily as we ought to.

## 6.04 How good is your cat's hearing?

Cats' other superpower (along with their sight) is hearing. Not only can they detect a wider range of frequencies than almost any mammal*, they can also accurately detect where a sound is coming from thanks to their independently swivelling ear flaps, known as pinnae.

Cats can hear sounds of a much higher pitch than humans – 64,000Hz (64kHz) compared with 20,000Hz (20kHz), about two octaves higher. This is particularly useful for locating mice, cats' favourite little snacks and/or playthings. **Mice and other rodents use high-pitched ultrasonic squeaks to communicate**, **which cats can not only detect but can even use to distinguish between different types of rodent**. Cats also have good sensitivity to lower notes, comparable to ours at around 20Hz. Most mammals' hearing focuses on one area of the scale but cats have a particularly large resonating chamber behind their eardrums that is divided into two interconnecting compartments, which increases their range.

Cats' pinnae can swivel 180° and are excellent hunting and climbing aids. They enable them to analyze sound in three dimensions and detect the source of noises made 1m (3ft) away to an accuracy of 8cm (3in). To do this, cats' brains evaluate the subtle differences between the sounds from their two ears in several ways. They judge lower-pitched sounds by the difference in synchronization (sound waves from a source hit one ear fractionally earlier than the other), and higher-pitched sounds by the difference in clarity (a sound will be slightly more muffled in the ear furthest from the source than in the other).

*The tiny list of mammals with a better hearing range than cats includes porpoises, ferrets and, weirdly, cows.

## 6.05 Do cats have a good sense of touch?

Touch, pressure, pain and temperature are the responsibility of the beautiful somatosensory system in humans as well as in cats. This is a network of sensors (also called receptors, nerve endings or sensory neurons) that create little electrical signals carrying information about pressure, heat, pain, vibration, smoothness, itchiness and more to the brain via axons (imagine these as minute cables connecting our touch receptors to our brains). Put your finger on your arm and the mechanoreceptors (sensors for touch) create a little electrical impulse that travels along your axons to your brain. It works the same way in cats.

Paws, claws and teeth are especially touch-sensitive in cats, but none more so than whiskers (or vibrissae), which are deeply embedded, tough, modified hairs packed with touch-sensitive mechanoreceptors at their base. They're extremely sensitive and most are found in patches either side of cats' noses – there are also smaller patches above each eye and, if you look closely, on the 'wrists' at the back of their forelegs.

**Cats can point their vibrissae forwards to provide close-up sensory information (helping make up for their eyes' inability to focus closer than 25cm/10in)**, and can also sweep them back for protection during a fight. They're sensitive enough to give cats detailed information about air movement and the objects they pass, as well as whether a gap is wide enough to crawl through.

## The Longest Whiskers

**The longest whiskers ever measured on a cat belonged to Missi, a Maine Coon from Iisvesi in Finland. They were 19cm (7.5in) long, according to *Guinness World Records*.**

# Chapter 07: Cat Talk

## 7.01 Why do cats meow?

Weirdly, meowing is specifically designed for communicating with humans and is rarely, if ever, used with other cats. Even more weirdly, very specific meows can have completely different meanings between different pairs of cats and owners. One cat's 'feed me' cry can be another's 'leave me alone' and it seems that usage develops symbiotically between each cat and owner to signify feelings such as hunger, annoyance, or the desire for attention, a stroke, or to have a door opened. **Cats and owners seem to train each other in a language unique to their pairing** – cat owners who were played meows from unfamiliar cats (yes, researchers really study these things) found it difficult to identify their meanings.

In 1944 the American psychologist Mildred Moelk studied cat vocabulary and identified 16 different meaningful cat–human and cat–cat voice signals. Her extraordinary work is still widely cited in studies today, and many biologists have expanded its scope. She divided cat signals into three groups: closed-mouth murmurs, vowel 'meows' starting with an open mouth that gradually closes, and strained open-mouth sounds that were the loudest and most urgent. Moelk even came up with a bizarre phonetic system to pronounce them, with a demand meow as 'mhrn-a':ou (try it out: it does make sense)*. The differences between the sounds are based on the duration of the cry, the fundamental pitch (the 'note') and whether the pitch changes during the cry. Moelk reckoned that cats have six different types of meow and attributed a broad association to each: friendliness, confidence, satisfaction, anger, fear and pain.

*A colon means that the preceding vowel is long.
An inverted comma indicates a stress.

## Murmurs: greetings or expressions of satisfaction

1. Purr ('hrn-rhn-'hrn-rhn)
2. Request or greeting 'chirrup' ('mhrn'hr'hrn)
3. Call ('mhrn)
4. Acknowledgment/confirmation ('mhng)

## Vowel meows: requests/complaints

These include the familiar meows, as well as sexual calls.

1. Demand ('mhrn-a':ou)
2. Begging demand ('mhrn-a:ou:)
3. Bewilderment ('maou?)
4. Complaint ('mhng-a:ou)
5. Mating cry – mild form ('mhrn-a:ou)
6. Anger wail (wa:ou:)

## Strained intensity non-meows: arousal, aggression or stress

There are many different versions of these cries.

1. Growl and anger wail
2. Snarl
3. Mating cry (intense form)
4. Pain scream
5. Refusal rasp – a type of hiss
6. Spitting

These strained-intensity sounds are mostly self-explanatory e.g. mothers growl at their kittens for transgressions. The hiss, though, is an expression of anger and if you don't back off, bro, you may be introduced to the spit.

## 7.02 Why do cats purr?

All cats purr* but we don't really understand why, because it occurs in a strange array of situations, both when cats are content and when they're stressed. There's even fascinating evidence that purring could help heal broken bones, but we'll come to that later.

Cats purr when they're anxious, when they're calm, when they're in pain, when they're giving birth, when they're injured, and when they want to be fed. Kittens start purring from the age of about one week while suckling. It's part of a reassurance signal between mother and kitten that is then adapted as the cat continues into adulthood.

Cats purr while both breathing in and breathing out, and although the noise sounds continuous to us, there is the tiniest pause between the two breaths. The purr comprises a series of fast beats, each made by the sudden separation of the vocal folds in the larynx as the glottis (the opening between the vocal cords) is closed and then opened. **These purr beats usually have a frequency of 20 to 40 times a second** (but can reach 100 times a second), and are controlled not by air passing over the vocal folds as with human speech, but by muscles contracting and releasing very fast. These are likely controlled by a free running neural oscillator (a mechanism in the cat's brain that generates a fast beat).

Things get more complicated by the fact that cats add meow-like intonations to their purrs, often when they want food. There are at least two different purrs: a 'normal' purr when the cat is not asking

*Well, all domestic cats purr, as do cheetahs. In fact all members of the cat family can either purr or roar, but never both. Lions and tigers have a sort of short splutter that sounds like a rubbish attempt at a purr. Doesn't count.

for anything and a 'solicitation' purr that humans find more urgent-sounding, less pleasant and harder to ignore, probably because the cat has added a tone in the 220–520Hz frequency range, close to the 300–600Hz range of a baby's cry. Yet another example of your cat manipulating you.

There's a theory that purring may improve healing and bone density in cats. Cats purr when recovering from injury or when visiting the vet, even when clearly anxious, and studies in humans show that certain vibration frequencies help heal fractured bones and surrounding muscles. The best frequency range is 25–50Hz for bone – similar to cats' most common purring frequency – and around 100Hz for skin and soft tissues. **So it's possible that purring in cats helps with physical repair – or at least to keep bones and tissues in good condition**. If there is such a connection, a purring cat is likely to be a marginally healthier cat.

## The Loudest Purr

**In 2015 Merlin, a rescue cat from Torquay in Devon, UK, registered a purr measuring 67.8 decibels, almost as loud as a dishwasher.**

## 7.03 Body language – what is your cat telling you?

### Tail talk

Tails are one of your cat's clearest communication tools, although you do need to know what you're looking for. **One common mistake of non-cat owners is to think that a cat's wagging tail means happiness, whereas it usually means the opposite**. If she's slowly flicking her tail, it tends to be a sign of irritation, and sometimes that she's about to lash out a sharp-clawed paw. Back off, she's telling you, and you'd be wise to take her advice. Similarly, a tail with its hair bristling and fluffed out wide is a sign of outright aggression.

On the other hand, if a cat comes towards you with her tail pointing vertically but relaxed in the air, then she's clearly feeling affection towards you – although we don't know if she's trying to signal that affection or just pointing her tail up because she is feeling affectionate. This is often followed by her rubbing her head against your legs.

### Eyes

If your cat slowly blinks, especially with semi-closed eyelids, she is feeling safe and content (slowly blinking back is a classic cat whisperer's trick). If she's not blinking, take a closer look at her pupils: if they are dilated, she may be excited or scared.

### Licking

My cat licks me awake pretty much every morning demanding to be stroked, and if the licking doesn't wake me, she'll give me a gentle pawing scratch with her claws half extended. Many cats lick their owners and this may be an echo of allogrooming (mutual grooming) that can continue for a surprisingly long time between mothers and kittens and between familiar cats, but its significance beyond affection and a demand for attention hasn't been established.

## Bunting

Cats can give a surprisingly strong head butt in order to get your attention for some quality petting time or before an expected feed, but they are also marking you with scent. **This head butting is known as bunting and is a clear sign of affection**.

Why a head butt? Cats have glands around their mouth, on either sides of their cheeks, along and around their tail and, crucially, on their forehead between eye and ear, and it's this part of the head they usually wallop you with. It's likely no coincidence that this is also most cats' favourite place to be stroked by their owners: that way, the contact contributes to scent marking as well as showing affection. We can't smell these scents, but they are important to your cat. Bunting is often accompanied by a relaxed ear position, slow, calm walking, and half-closed eyelids.

## Anus-in-face

Every cat I've ever had has thoroughly enjoyed pointing its anus in my face, often at revoltingly close proximity. I've certainly seen my cats' anuses hundreds more times than I've seen my own. Actually I don't think I've ever seen my own. Have you? Anyway, I always wondered if anus-display was a sign of something: disdain perhaps? Power? Sheer joie de vivre? Biologists have little to say on the subject, but it's thought that your cat would only turn its back on you if he trusted you.

## 7.04 How do cats talk to each other?

Cats are usually solitary but there are exceptions. Lions can sustain large, high-functioning prides of males and females, feral cats (basically ex-domestic cats) can live together in large colonies, and cats from the same litter often tolerate each other pretty well. Unrelated cats can sometimes live together in harmony if they're introduced at an early age*, and even urban cats find it difficult to avoid each other entirely, so have had to develop ways of communicating to get along and avoid fights.

Cats rarely talk to each other at all. Meowing is reserved almost exclusively for communicating with humans (see p93), and howling or yowling is only used for stand-offs and fights, which are rare. Instead, they employ many of the same body-language signals they use with humans, along with a complex system of tail displays, rubbing, exchanging scents, and especially allogrooming – mutual grooming that can continue for a surprisingly long time.

### Tails

**Cats raise their tails straight in the air with a relaxed appearance if they are happy to approach each other** – although it's unclear if this is a deliberate peace sign to communicate or just happens in this situation. On the other hand, if a cat's tail is raised, flicking to either side and bristling, it's a sign of fear or aggression and usually accompanied by other angry signals.

*Allowing a new kitten into the family to join my lovely, soppy 10-year-old tabby Tom was, however, an unmitigated disaster.

## Eyes

Cats that are happy to tolerate each other's company will often blink slowly with semi-closed eyelids, as they do with humans. Cats are much more aware of whether other cats' pupils are dilated, which can indicate excitement or fear. Long stares can be a sign of aggression.

## Ears

**Cats' ears are controlled by over 20 muscles and can swivel 180°**. When pointing up and forward, they may indicate a cat is happy and ready to play. When erect and curled backwards, they're likely to signify aggression. When defensive, cats lay their ears flat out, pointing sideways or sometimes backwards, although this can also be an invitation to play.

## Rubbing

Cats don't seem to use bunting (see p98) with other cats in the same way they do with humans, but if two cats are happy to meet, they'll rub, which also allows them to exchange scents. Feral cats use this signal in colonies, although again it's not clear if they do this in order to invite friendship or as a result of the friendship itself.

## Allogrooming

When familiar cats meet they often groom each other. The practice may relate back to the bonding experience they had when being groomed by their mothers. They fastidiously lick each other, exchanging scent at the same time, and the result does seem to be a decrease in conflict. It may be that a group of cats exchanging scent helps create a crowd-sourced communal scent that strengthens bonds. Weirdly, it's usually the dominant, most aggressive cats that do most of the grooming.

## Fur

The follicles at the base of cats' hairs can be used to straighten those hairs in a process called piloerection. This is triggered by the automatic release of adrenaline when a cat is scared, which causes the bristling effect. This puffs the cat up to look as large and threatening as possible (although it's not clear that the cat knows or wants to achieve this), so it's used when the cat is feeling both aggressive and defensive at the same time.

## Full body arch

The full body arch with piloerect fur prickled up like a porcupine is a clear sign of aggression – my cat will instantly fizz up like this when my dog wanders past, only to curl out of it once he's gone. I don't know about you, but it always scares the living daylights out of me. An arch without the bristling fur can occasionally be part of a 'stroke me' signal towards humans.

## Belly roll

The brilliant magazine *Annals of Improbable Research* unearthed a 1994 study entitled 'Domestic Cats and Passive Submission'. One Hilary N Feldman spent six months observing 175 semi-feral cat rolls and noted that 138 of them 'had an obvious recipient'. Females rolled mostly while in heat, but males mostly did so 'as a form of subordinate behaviour'. On the other hand, cats roll on their backs to oppose a violent attack, bringing up their powerful back legs with fully unsheathed claws, ready to kick with a vicious blow. Although mercifully rare, fights like this are horrible to watch and can result in nasty injuries (see p52).

# Chapter 08: Cats vs Humans

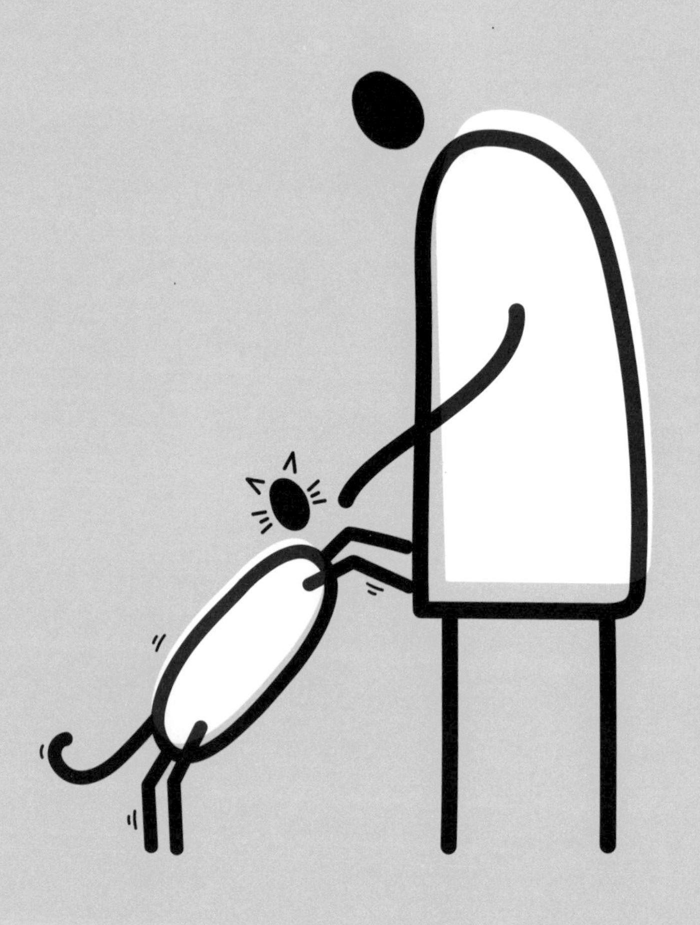

## 8.01 Why do we love cats rather than ferrets?

Cats are obstinate, aloof, fickle, demanding and, after vomiting on the stairs and bringing fleas and dead animals into your house, prone to buggering off to live with the neighbours. They are notoriously difficult to train, and even if you can train them, it's hard to get them to do anything helpful. On the other hand, ferrets are intelligent, playful, useful, adaptable and can be trained to hunt rodents and, crucially, rabbits – the scourge of farmers across the world. That alone should put them above cats in the pet pecking order. They also sleep a lot, can use litter trays and enjoy the company of humans. And the collective noun for them is a 'business' of ferrets. What's not to love?

African wildcats were likely first welcomed into human homes soon after we began farming because they were handy for catching the small tasty rodents attracted to our grain stores. After all the effort of inventing agriculture, the last thing our ancestors wanted was for mice to come along and blow it all. But, now that cats' hunting skills are relatively useless (unless you own a grain silo), why are they still hanging around the house?

Well, partly because ferrets also like to escape, so need to be caged and require constant supervision when let out. They're also prone to stealing and hiding your things, as well as eating pretty much anything that will fit in their mouths. They're vulnerable to a range of health problems, too, so vet bills are likely to be expensive. Unless you're a farmer (or just really hate rabbits) ferrets are really only as useful as cats.

The evolutionary masterstroke that pushed cats ahead wasn't so much their mouse-catching as their ability to tolerate us, and to look cute. They have one big advantage, which relates more to human psychology than anything else: they're much easier to anthropomorphize than ferrets. **Cats' facial structure is remarkably similar to that of young humans: both have flat faces, high foreheads, small noses and large, forward-facing eyes**. This has made us think (invariably wrongly) that we can relate to them. Combine this with the facts that cats are relatively easy to care for and independent enough to be left at home all day, and you have a companion animal that can wrap itself around a modern urban human, satisfying our deep parental urge, and distracting us from the pointlessness of existence.

## *Legendary Cats*

### Tarawood Antigone

**World records for cats should be taken with a pinch of salt, but apparently in August 1970 this Burmese gave birth to 19 kittens in Oxfordshire, UK, of which 15 survived: 14 males and one female.**

## 8.02 Are cats good for your health?

Our cats are good for our health, aren't they? Everyone says so, and it makes such a lot of sense: caring for another creature keeps us alert, their companionship makes us happy, and what could be wrong with all that furry cuteness? Well ... there's good news and bad news.

### The good news

Yes, there is a study showing that cat ownership equates with good health including decreased risk of death from all cardiovascular diseases and better survival rates from heart attacks. Another investigation concluded that sleeping with your pet may benefit rest (although a significant number of subjects also found that their pets disrupted their sleep) and there is also research indicating that children brought up with pets are less susceptible to asthma. An excellent Swedish study looked at people with heart problems and found that those who owned dogs had better health outcomes than those who didn't, while an Australian study found that **children with a cat or dog in the house were 30% less likely to have symptoms of gastroenteritis than those with no pets**.

And then there are the surveys. One found that 87% of people felt owning a cat had a positive effect on their wellbeing and 76% thought the company of their cats helped them better cope with everyday life. Another survey found that owning a cat has a small beneficial effect on how attractive you are – women cat owners were perceived as 1.8% more attractive and men cat owners 3.4% more (male puppy owners, meanwhile, were seen as 13.4% more attractive). But beware: surveys are tricky because they ask people their opinions, which most scientific studies discount as highly unreliable – scientists prefer to design clever ways to exclude opinion and concentrate on evidence. What's more, surveys aren't peer-reviewed and are sometimes simply PR dressed up to look like important research in order to flog something or get attention. So treat them with caution.

## The bad news

On the flip side, I'm sorry to report that many academic studies also show that owning pets has no effect on health, and can even be associated with negative effects. A 2005 study from the Australian National University in Canberra reported that **adults aged 60–64 living with pets are more likely to be depressed, have poor mental health, suffer higher levels of psychoticism, take pain medication and be in worse physical shape than pet-free people**. Another Australian study showed that pet ownership had no effect on elderly people's physical or mental health. Researchers in Finland found that pet ownership is associated with poorer rather than better perceived health as well as higher BMI, high blood pressure, kidney disease, arthritis, sciatica, migraines and panic disorders. Research from Queen's University Belfast showed that pet owners with chronic fatigue syndrome (CFS) were convinced that their pets gave them many psychological and physical benefits, but in fact were just as tired, depressed and worried as CFS sufferers without pets.

Why, though, do we rarely hear these stories? Partly because people want to think that cats are good for us (we do love them, after all), but also because of positive-outcome publication bias. Positive study results are much more likely to be published (and cited in other studies) than non-positive results, despite the fact that science dictates that a theory proven wrong is just as valuable as one proven right. Another factor is the epidemic of terrible science journalism. I spent two depressing weeks trying to track down the sources of numerous articles with titles such as '10 scientific reasons why having a cat is good for you' and discovered that less than a third of the 'scientific reasons' given were attached to any academic research. The rest tended to be blindly regurgitated opinions, assumptions and even clear falsehoods peddled as fact. Many claims, such as cats' ability to help children with autism, weren't borne out when you read the cited study, which, in this case, was specific to dogs. I don't want to pour water on anyone's furry love-fire and, God knows, I love my cat, but let's stick to the facts, shall we?

## *Legendary Cats*

### Tara

**You could lose a large proportion of your life looking at cat videos on YouTube, but this one is a great, short watch. In 2015 CCTV caught Tara saving her owner's four-year-old son from a nasty attack by a neighbour's dog, racing to his rescue and attacking the dog as it tore at the boy's leg, before chasing it away. Tara swiftly became a hero-celebrity, whereas the dog was certified dangerous and euthanized shortly afterwards. Search 'My Cat Saved My Son'.**

## 8.03 Are cats BAD for your health?

Let's talk zoonoses. Zoonoses have nothing to do with noses. They are infectious diseases that can spread from animals to humans. Rabies, ebola, SARS and coronaviruses are all zoonoses, as is toxoplasmosis, caused by a protozoan parasite found in 30–40% of domestic cats that can infect humans through contact with feline poo.

The strangest thing about toxoplasmosis is that it can make animals, including humans, more reckless. A team of researchers from the University of Geneva found that when the *Toxoplasma gondii* parasite infects rodents they become more daring: drastically less afraid of cats and less fearful in general (good for cats who like hunting mice, less good for mice who like being alive*). It even made the mice attracted to the scent of cats' urine. Evolutionary biologist Jaroslav Flegr then went on to spend many years studying toxoplasmosis' effects on human behaviour, and he discovered that infected men are more likely to disregard rules, be excessively suspicious or jealous, and have significantly delayed reactions. **When he studied drivers and pedestrians injured on the roads in the Czech Republic, he found they were more than twice as likely to be infected with *Toxoplasma gondii*.**

Having toxoplasmosis doesn't make you special – around half the world's population may be infected already but the vast majority show no signs of the disease. However, the sheer number of people infected means that the small proportion who do, suffering flu-like symptoms, seizures and eye issues, adds up to a big problem. The disease is particularly dangerous for those with compromised immune systems, and pregnant women with acute toxoplasmosis can also infect their babies. The good news for cat-lovers is that cats are less likely to infect you than poorly cooked meat containing toxoplasmosis cysts (dormant microorganisms). Nonetheless, pregnant women should avoid contact with cat litter in case it's infected.

Weirdly, cysts (dormant microorganisms) caused by *Toxoplasma gondii* are found in particularly high concentrations in the brain regions of infected mice that process visual information, and cause inflamed neural tissue throughout the brain. More research is planned to examine how this neuroinflammation alters various behavioural traits, but it seems that *Toxoplasma gondii* may have co-evolved to cats' benefit: the parasites in cat poo infect mice and affect their sight, making them easier for cats to catch. Genius.

On top of this, research shows that cats can trigger eczema in children and, of course, cats bite, inflicting around 400,000 every year in the USA alone. Many of these bites become infected with the bacterial infection *Pasteurella multocida*, which appears around 12 hours later. There's also a connection between cat ownership and depression: David A Hanauer at the University of Michigan Medical School found that an extraordinary 41% of patients with cat bites suffered from depression compared with 9% of the research group as a whole. Then there's the imaginatively named cat scratch disease (CSD), caused by the *Bartonella henselae* bacterium, and creeping eruption (AKA cutaneous larva migrans), caused by *Ancylostoma braziliense* (cat hookworm). Once again, avoiding cat poo is your best bet.

*It turns out that the cysts (dormant microorganisms) caused by Toxoplasma are found in particularly high concentrations in regions of infected mice brains that process visual information, and cause inflamed neural tissue throughout the brain. More research is planned to examine how various behavioural traits are altered by neuroinflammation, but it seems that Toxoplasma gondii may have co-evolved to cats' benefit.

## 8.04 How much does a cat cost to run?

Battersea Dogs and Cats Home estimates **the typical cost of owning a cat in the UK at around £1,000 per year – or £18,000 over the course of its 18-year lifetime** (it's quite optimistic about the average cat's lifespan). The American Society for the Prevention of Cruelty to Animals estimates costs at only $634 ($11,412 over 18 years) but both of these figures exclude the cost of buying your cat. It's a huge amount of money, but even so, cats are cheaper than dogs, which cost £445–£1,620 per year in the UK and $650–$2,115 in the USA – that's a total of £5,785–£21,060 or $8,450–$27,495 over an average 13-year lifetime to run, according to The Dog People (again, before purchase costs are factored in).

Of course, the true cost depends on how much you want to spend. A pedigree cat can easily cost up to £1,000 ($1,400) to buy, with higher pet insurance costs and grooming fees along with it, compared to a rescue moggy like mine, who cost me a donation of around £70 ($95). Food costs are likely to be your biggest expenditure and vary wildly from £160 to £2,000 ($225 to $2,800) a year in the UK depending on the brand you buy and your cat's needs (cats with specific nutritional requirements may need more expensive food). Another important cost is pet sitting – who is going to look after Mr Fluffs when you're away? In the UK professional pet sitters and housing cats in catteries can easily cost an extra £1,000 ($1,400) a year, though you might be lucky enough to have a neighbour who'll help you out of pure love.

Other costs are regular vet's bills for vaccinations and check-ups, setup costs such as microchipping and installing a cat flap, and kit:

bowls, litter and litter trays, toys and a carrier for trips to the vet. And you MUST insure your cat. I didn't notice that the insurance on my beloved Tom Gates had lapsed and his healthcare in the last year of his life cost me a crippling £3,000 ($4,100). Cat insurance costs £35–£300 a year in the UK and $300–$900 in the USA depending on the policy and where you live (it's more expensive in big cities), but costs can shoot up for older cats. Many companies simply won't insure cats over eight or 10 years old.

## *Legendary Cats*

### Tommaso

**This black cat inherited $13 million from his owner, Italian property magnate Maria Assunta, who died aged 94 with no close relatives. Maria specified that the fortune should either go to Tommaso, who had wandered into her home as a stray, or to an animal welfare charity that would look after him, but since she was unable to find one that she was happy with, after her death, the money was held in trust by Assunta's nurse on his behalf.**

## 8.05 Why does your cat bring you presents of dead animals?

Most domestic cats are well-fed on a never-ending supply of nutritious-yet-tasty meals from packages covered in photos of the cutest ickle kitties you ever did see. So why does your cute ickle kitty leave you presents of half-eviscerated rodents at the top of the stairs?

The standard answers to this used to be:

a. Your cat sees you as a big, lazy, cat-baby who's rubbish at hunting and thinks you need feeding.
b. Your cat is trying to teach you to hunt.
c. Your cat wants you to be ever-so-proud of his hunting abilities.
d. It's a present – stop being so ungrateful and say thank you.

There's little evidence, however, that any of these are true. The most likely answer is that your cat is a natural predator driven to hunt and kill whether you've fed him his expensive vet-assured meal of duck-in-gravy or not. And although mother cats do often bring their kittens dead mice to eat, there's no evidence that your cat sees you as its kitten. **He probably brings his prey home as part of the hunting impulse, but the evolutionary imperative evaporates when he arrives back and can't be bothered to eat it**. He'll drop it randomly when he gets distracted – possibly by your presence. It may look as though our cats are offering us presents, but that's because we're so desperate for evidence that they love us, we ignore the painful truth.

## 8.06 Can you take your cat out on a lead?

**A wide range of leads and harnesses is available if you fancy taking your cat out for a walk**, but although some animal trainers encourage it, the UK's Royal Society for the Prevention of Cruelty to Animals (RSPCA) thinks it's a pretty bad idea. Cats are highly territorial and find new and changing environments stressful, it says. However, it stops short of advocating a ban on all cat leads. 'All we want cat owners to consider is that every cat is an individual,' says Dr Samantha Gaines, head of the RSPCA's companion animals department. 'For some, walking on a lead may be suitable, but we need to be careful that we're not just thinking of cats as dogs.'

Cats fiercely value their freedom and sense of control – something you take away from them the moment you put them on a lead. 'Taking steps to provide an indoor environment which has plenty of opportunities to be active and mentally stimulated is likely to be more beneficial for the cat's welfare than walking them on a lead,' Gaines says.

### A Kindle of Kittens

**There are lots of collective nouns for cats, including a 'clowder', a 'clutter', and a 'glaring'. A group of kittens is called a litter or a 'kindle'.**

## 8.07 Why are people allergic to cats?

A surprising **10–20% of the world's population are thought to be allergic to domestic pets, and cat allergies are twice as common as dog allergies**. An allergy is a hypersensitivity of the immune system to substances that are usually harmless, and the most common reactions are itchy eyes, coughing, sneezing, nasal congestion and rashes. Some people, however, can develop allergic asthma or rhinitis, and the worst cases of these can be fatal.

Sufferers often think that cat fur is to blame but the culprit is almost certainly one of eight proteins found in cats' saliva, anal gland excretions, urine and, especially, the oily sebum from hair follicles that's spread via microscopic flakes of dander (cat dandruff).

The most problematic allergen by far is the jauntily named Fel d 1 protein (the others are named Fel d 2 to Fel d 8), which is found in cat saliva and dander and causes 96% of cat allergies. If you're allergic, exposure to Fel d 1 causes plasma cells in your blood to produce Immunoglobulin G or Immunoglobulin B antibodies designed to bind to allergens and neutralize them (despite the fact that the allergens are otherwise harmless). This triggers the release of inflammatory chemicals such as histamine, which are supposed to help white blood cells and proteins tackle the suspected pathogen. However, people with allergies end up overproducing them, leading to itching and tissue swelling.

If you're allergic, you may be able to tackle the problem by taking antihistamines, regularly washing bedding, vacuuming and bathing your cat (good luck with that!). Alternatively, try getting a hypoallergenic cat that doesn't shed much hair or has lower Fel d 1 levels.

## 8.08 Why do cats love people who hate cats?

Cat-haters, ailurophobes (people with a persistent, excessive fear of cats) and those allergic to cats often say that cats are more attracted to them than ailurophiles (people who love cats). In his book *Catwatching*, Desmond Morris suggests that cats are drawn to people who don't try to stroke them, and that conversely, people who love cats look at them more intently, which can make them anxious. It's a nicely counterintuitive argument.

More recently John Bradshaw (renowned cat behaviour expert and director of the Anthrozoology Institute at the University of Bristol) tested this theory using people who either liked or were repelled by cats and found the opposite to be true: seven of the eight cats he used avoided the cat-phobics. Meanwhile, the single cat that bucked the trend jumped on the cat-phobics' laps purring loudly. Bradshaw suspected that on the few occasions when cats do prefer cat-phobics, it makes such an impression on them that they come away thinking it happens all the time.

## 8.09 Why are cats so difficult to train?

Unlike dogs who benefit hugely from human interaction, cats aren't particularly interested in making us happy. They are solitary ambush predators who get together for mating or kitten-rearing and that's pretty much it. They simply don't have the motivation for working together and hence are hard to train. Frankly, it's a miracle that we get them to enjoy our company at all, and if they weren't such natural rodent-slayers, it's unlikely they would ever have made it into our homes.

**Although many cats show us affection, they don't seem to need or crave our approval**, and nor are they especially food-motivated, all of which makes them notoriously difficult to train. That said, it's easy to get cats to use a litter tray – so easy that they'll often train themselves – to use a cat flap, to return to your house, and to come when called. And it's not impossible to train cats to do other things as well, which is why there is a multitude of books claiming to help you*.

Cat training usually starts with food rewards, and then shifts to the use of a clicker. To begin with, the clicker is used at the same time as the food reward, and in time the cat comes to consider the click itself as the reward. It's a classic conditioning technique called secondary reinforcement, and it works well as long as you regularly intersperse clicker-only sessions with clicker-and-food sessions to retain motivation.

*There is one excellent book called *The Trainable Cat* by John Bradshaw and Sarah Ellis, but it focuses on training your cat to be less anxious, antisocial and stressed out by newcomers, visitors or trips to the vet rather than to jump through fiery hoops.

**If you have trained your cat to perform any of these tricks, please be aware that I am merely desperately jealous. I tried so hard. So bloody hard.

So, what can you train your cat to do? Common tricks include training her to sit, to jump over obstacles, to offer her paw to be shaken, to give a high-five, to walk on a lead and to jump through a hoop on to a target. Which really begs the question: why?**

## Most Cat Tricks Performed in One Minute

**In February 2016 a cat called Didga performed 24 tricks in one minute with her owner Robert Dollwet in Tweed Heads, Australia. Tricks included jumping, high-fiving and riding a skateboard, according to *Guinness World Records*.**

## 8.10 Is it fair to keep your cat inside?

It's a highly contentious question: should you keep your cat locked inside or allow her to roam at will? Insiders are baffled that Outsiders would even dream of letting their beautiful Woo-Woo out to fight for her life against cars, dogs and Marmalade 'The Slayer' Jones from five doors down. It's true that outdoor cats are thought to kill lots of wildlife and have shorter life spans (because of the greater risks they face outside), but Outsiders argue that at least their cat is Living Her Best Life, doing what comes naturally and hanging out in the fresh air.

The truth is that it's perfectly possible to keep your cat inside. **Cats are solitary and aren't designed to hang out with other cats, so the to-ing and fro-ing of cats in local gardens is likely to cause more anxiety than friendship**. But they do have a natural instinct to hunt and climb, along with an appetite designed to support that activity, so indoor cats have a few specific requirements.

The kit list is pretty obvious: litter trays, at least one scratching post, bowls, and spots to hide and hang out, preferably some of them up high. Puzzle feeder devices that cats have to explore and play with to extract the food are useful, too. Much more important, though, is your time and attention. Inactivity can lead to boredom, stress and obesity, and owners are an indoor cat's main distraction. They need lots of stimulation with games, grooming, stroking and scratching together, and chasing and hunting toys – balls, feathers, cardboard boxes and mouse-alike thingies. And you. Lots of you.

## 8.11 Why do cats massage soft things – like your lap?

Most cats will massage or knead anything soft, commonly with eyes half-closed in a clear state of bliss. They press down with alternate paws at intervals of one to two seconds, often directly on their owners. As they do so, they also stretch their toes and extend their claws, which can get caught in whichever material they're massaging. While cats seem to enjoy this, it can be particularly painful if they're on your lap, giving rise to the classic cat owner's conundrum: maintain the fragile bond between cat and human or just get those digging claws out of your flippin' crotch?

This kneading behaviour is most common in kittens but many cats continue to knead into adulthood, usually when feeling content and unthreatened. It tends to be accompanied by a clear purring and sometimes, as an extra sign of pleasure, uncontrolled dribbling (my cat also dribbles copiously when I stroke her, but that's probably because I am exceedingly good at stroking).

Biologists think **kneading may be a behavioural remnant from cats' kitten days** when they would knead their mother's teats to stimulate milk production. As feeding gave them pleasure, they link the behaviour with a positive experience. Cats may now have displaced or borrowed this kitten-to-mother communication to show similar affection to their owners, which is why your cat kneads your lap. My cat's uncontrollable dribbling may also be related, triggered in a phantom anticipation of her mother producing milk. On the other hand, kneading is often followed by sleeping, which strengthens a completely different theory that it's an evolutionary throwback to wildcats' habit of pressing foliage down to make a temporary nest.

## 8.12 What's the climate impact of a cat?

It's lovely owning a cat, but they do place a burden on the environment in the form of the faeces they produce and the food they eat, which requires energy to produce, harvest and transport. In *Time to Eat the Dog?*, authors Robert and Brenda Vale estimate that the eco-footprint of a cat equates to that of a Volkswagen Golf driven 10,000km (6,200 miles) a year, and that a cat's ecological footprint is about 0.15 hectares (0.37 acres) per year compared to a dog's 0.84 hectares (2.08 acres).

A 2017 UCLA study concluded that US dogs and cats consume about 19% of the amount of dietary energy that humans do – adding the equivalent burden of an extra 62 million people. They also produce lots of poo – another 30% of the amount that humans do. **Together, food for dogs and cats is responsible for 25–30% of all the environmental impacts from animal production in terms of land, water, fossil fuel, phosphate and biocide use**. The study acknowledges that pet food is invariably made from meat by-products that are not usually consumed by humans, but counters that if dogs can eat those, humans should also be able to. Admittedly, not many of us enjoy tripe, lungs and entrails on a regular basis right now, so doing so would require a big cultural shift. Thankfully, though, they can be delicious (I'm particularly partial to a bit of bone marrow).

The study says that: 'People love their pets. They provide a host of real and perceived benefits to people …'. Nevertheless, we should be aware that our pets represent a significant ecological burden, which we must take into account when mitigating our own impact. This opens up a world of moral and ecological relativity where we have to balance unquantifiable emotional impact (I love my cat so much) with quantifiable climate impacts (my cat eats another 19% of my dietary energy requirements), and this can lead us into difficult territory. After all, one of the most effective ways of cutting your $CO_2$-equivalent emissions is to cut down on the amount of children you have: having one fewer child saves 58.6 tonnes (64.6 tons) of $CO_2$-equivalent emissions per year (changing to a plant-based diet only saves 0.8 tonnes (0.9 tons) of $CO_2$-equivalent emissions per year). Of course, we love our kids, and it's both impossible and horrific to quantify whether or not all that extra love in the room outweighs the downsides. There is definitely a balance to be achieved, and discussions to have, but is it just a short jump from cutting down on family pets to a one-child policy?

## 8.13 Are cats bird-murdering machines?

Cats hunt and kill wildlife, including birds, although whether or not they have a more negative impact than other predators, such as magpies, rats, foxes or coyotes (or indeed whether removing rat-catching cats from the world would be of benefit) is hotly debated.

The Mammal Society caused a media furore in 1997 when it estimated that 275 million animals a year are killed by pet cats in the UK, a figure based on forms completed by their youth wing and extrapolated from a survey of 696 cats. There has been a fair amount of controversy about the accuracy of the figures, but there's little doubt that cats eat other animals. A 2013 study published in *Nature* estimates that cats kill 1.3–4 billion birds and 6.3–22.3 billion mammals in the USA every year.

So, are cats bad for birds? Well, it's not that simple. **The Royal Society for the Protection of Birds (RSPB) says that cats catch an estimated 27 million birds in the UK every year, but also that 'there's no clear scientific evidence that such mortality is causing bird populations to decline'**. There is evidence that cats mostly take weak or sickly birds, it notes, saying it's 'likely that most of the birds killed by cats would have died anyway from other causes before the next breeding season, so cats are unlikely to have a major impact on populations.'

The worst impacts to animal populations are seen on islands that have never had cat-like predators before. There, cats can devastate local wildlife that simply hasn't developed the means to protect itself. Some of the strongest opposition to cats comes from Australia and New Zealand, where species of small marsupial and flightless bird have become extinct (although it's unclear whether this is entirely due to cat predation) and several municipalities have strict cat restrictions – ranging from confining them to houses to bans on ownership in new suburbs. But there's no clear evidence that any of these restrictions help wildlife, and sometimes the data suggests the opposite, possibly because cats also prey on rats, which in turn prey on birds.

Of course, domestic cats are not the only killers of birds or eggs – feral cats, foxes, magpies, rats, birds of prey, starvation and a simple failure to thrive kill the vast majority of wildlife. Although feral cat populations are supported by people keeping domestic cats (and then losing them), their impact on wildlife isn't absolutely clear. Meanwhile, in *Cat Sense* John Bradshaw notes that 'the UK has at least ten brown rats for every cat'. Rats are well-known to negatively impact bird and small mammal populations, so the anti-cat lobby should be careful what it wishes for.

# Chapter 09:
# Cats vs Dogs

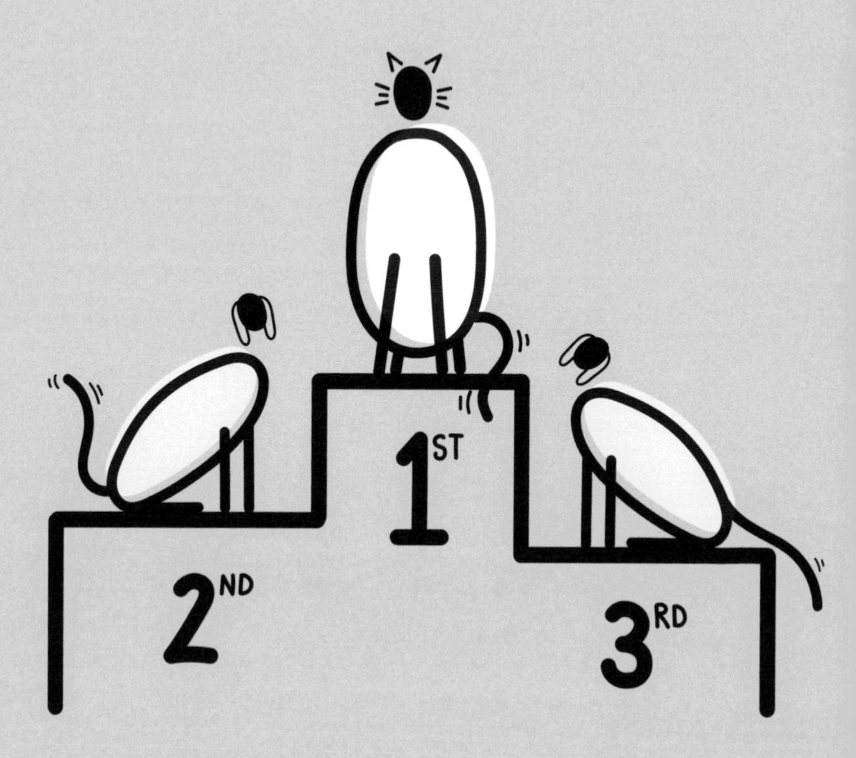

## 9.01 Can one species be better than another?

Before diving headlong into the great Cats vs Dogs debate, let's pause for a conceptual biology moment. Don't worry – this shouldn't hurt too much.

With our opposable thumbs, powers of abstract thought and fabulous music taste, we humans like to think we're superior to all the other species on Earth. Apes and dolphins may not be far behind, but earthworms and plankton? Pah! Look at what we've achieved: our impact on the Earth is so great that the Holocene era (the 12,000 years since the last ice age during which human civilization developed) is now thought to be over, replaced by the Anthropocene, an era defined by humans' pre-eminent impact on the planet. Following the invention of the spork, selfie stick, and Justin Bieber, it's fair to say that species don't come much more perfectly formed than us. Yeah! Go, humans! Except, of course, the Anthropocene is defined by disastrous markers starting with radioactive pollution in the 1950s, a striking acceleration of $CO_2$ emissions, mass deforestation, biodegradation, war, inequality and the global mass extinction of species.

On the other hand, earthworms' ancestors survived five extinction events and existed for 600 million years compared to humans' paltry 200,000 years. Darwin thought that earthworms played one of the biggest starring roles in the history of the world, ploughing and fertilizing our soil and making it possible for us to grow food. And plankton? Well, just look at the numbers: 7.8 billion humans is pretty paltry compared to the SAR11 plankton population at $2.4 \times 10^{28}$. That's 24,000,000,000,000,000,000,000,000,000 plankton to you, squire.

So, **asking whether cats are better than dogs is generally considered a fool's game, a bit like asking 'What's better: a tree or a whale?'** A tree excels at being a tree and a whale excels at being a whale. An earthworm isn't better or worse than a human – it's excellent at being a terrestrial hermaphroditic invertebrate that respires through its skin and lives underground. And even then, species are thought never to be at an evolutionary optimum, but always in some form of adaptation in relation to their situation. Cats' and dogs' domestication is particularly interesting: in evolutionary terms they are wild hunting predators that have only relatively recently moved into our houses, so are probably only at the start of an adaptation phase. Check in again in half a million years and they may be biologically very different. And looking at the way the Anthropocene's going, their beloved humans may no longer exist at all.

## 9.02 Cats vs Dogs: social and medical

The previous pages went to great lengths to explain why comparing cats with dogs goes against conceptual biological principles. But where's the fun in that? Come on, let's play Cats vs Dogs!

### Popularity

Dogs are way more popular than cats in the UK* (although statistics do vary wildly).

23% of households own at least one dog

16% of households own at least one cat

**Winner: DOGS**

### Love

Owners of both species are pretty wild about their pets, but which animal loves us more? Neuroscientist Dr Paul Zak analyzed saliva samples from dogs and cats to find out which contained more oxytocin (the hormone associated with love and attachment) after playing with their owners (see p63). Cats' oxytocin levels increased by an average of 12% but dogs' rose by a whopping 57.2%. That's a six-times greater increase. Dr Zak even twisted the knife, noting: 'It was a nice surprise to discover that cats produce any [oxytocin] at all.'

**Winner: DOGS**

*Pet Food Manufacturers' Association Pet Population 2020 report

## Intelligence

Dogs' brains, at an average 62g (2oz), are bigger than those of cats', which average 25g (0.9oz). But that doesn't necessarily make them cleverer – sperm whales have brains six times the size of humans', but are still considered less intelligent because, among mammals, we have the largest cerebral cortex (the area responsible for information processing, perception, sensation, communication, thought, language and memory) relative to the size of our brains. Another measure of intelligence is the number of neurons in an animal's cerebral cortex. Neurons are fascinating because they have a high metabolic cost (they use a lot of energy to keep running) so the more neurons we have, the more food we need to consume, and the more metabolic machinery we have to run to turn that into usable fuel. Because of this, each species only has as many neurons as they absolutely need, and a paper published in *Frontiers in Neuroanatomy* found that **dogs have more neurons in their cerebral cortex than cats** – about 528 million compared with 250 million. Humans trump them both with 16 billion, though. The researcher who developed the measuring method said 'I believe the absolute number of neurons an animal has, especially in the cerebral cortex, determines the richness of their internal mental state ... dogs have the biological capability of doing much more complex and flexible things with their lives than cats can.'

The brain issue really depends on what's most important to a particular animal – dogs are pack animals, so need more communication skills to function together, and these are centred in the frontal lobe and temporal lobe, whereas cats are solitary hunters and may need more motor function skills, centred in the frontal lobe's motor cortex, to control escape abilities such as climbing.

**Winner: DOGS**

## Convenience

Cats are cheaper to buy, keep, feed and care for. They are independent, don't need walking, and can be left alone for much longer than dogs. They will happily poo and pee outside, and usually not in your own garden (good for you, not so good for your neighbours). And dogs? Dogs are not convenient.

**Winner: CATS**

## Sociability

Cats are solitary and territorial animals but get physiological benefits from contact with humans. Dogs are sociable with each other, although prefer the company of humans. They respond to many human commands and requests, and enjoy physical contact – which, like with cats, provides physiological benefits.

**Winner: DOGS**

## Eco-friendliness

Cats kill millions of birds every year (although the impact and exact number are strongly disputed) and both dogs and cats may reduce biodiversity. Dogs, on the other hand, have a larger ecological footprint: feeding a medium-sized dog requires 0.84 hectares (2.08 acres) of land a year compared to 0.15 hectares (0.37 acres) for a cat.

**Winner: CATS (just)**

## Health benefits

**Both dog and cat owners get positive hormonal benefits from contact with their pets (which helps lower stress levels) and have better immunoglobulin levels than non-owners**, potentially offering higher protection from gastrointestinal, respiratory and urinary tract infections. However, many of the grander health claims associated with pet ownership have been questioned in more recent studies. Dog owners tend to exercise more than cat owners and non-pet-owners, which is likely to lower cardiovascular risks and increase survival rates after heart attacks. But these benefits are countered by the facts that every year in the UK 250,000 people have to visit minor injury and emergency units after suffering dog bites and two to three people die as a result of dog attacks. Globally, rabid dogs cause the death of around 59,000 people a year, according to the World Health Organization.

**Winner: CATS**

## Trainability

Dogs: the average dog can be trained to memorize 165 words and actions, catch a ball, sit, give a paw, jump, heel, lie on their bed, roll over, wait, and – at a push – stop humping Auntie Gladys's leg.

Cats: HAHAHAHAHAHAHA.

**Winner: DOGS**

## Usefulness

Mouse-hunting is extremely useful – to the handful of people who own grain silos/farms or have a mouse infestation. To the rest of us, it's a bit irritating. Bird-hunting, on the other hand, is downright out of order. And that's about all cats offer – other than the obvious joy they give us whenever they can be bothered to seek our attention. Dogs, on the other hand, are useful for hunting, sniffing out contraband and explosives, tracking in the wilderness, diagnosing diseases, rescuing lost or trapped people, guiding the visually impaired, herding sheep, guarding houses … I'll stop now – you know what I'm saying.

**Winner: DOGS**

## 9.03 Cats vs Dogs: physical head-to-head

### Speed

Cheetahs are the fastest animals on land, able to run at 117.5km/h (73mph). But your cat, bless him, isn't a cheetah. If he can be bothered, he's likely to be able to achieve 32–48km/h (20–30mph) for short bursts. This compares shabbily with greyhounds' maximum speed of 72km/h (45mph), but pretty well against a lumbering Golden Retriever at 30km/h (19mph).

**Winner: DOGS**

### Stamina

Dogs are hands-down winners here. Cats are ambush predators, capable of patiently stalking their prey for hours before sprint-pouncing. Dogs aren't built for sprinting but for long-range aerobic endurance chasing (much like myself, as it happens). Humans have co-opted this ability for travelling across ice and snow, and the stamina that sled dogs show is extraordinary – animals in the Iditarod Trail Sled Dog Race cover 1,510km (940 miles) of sparsely populated Alaska in eight to 15 days.

**Winner: DOGS**

### Hunting ability

Despite being fed regular meals, almost all domestic cats retain both the urge and the skills to hunt, often bringing home mice and birds in differing states of completeness. Conversely, most dogs have a chase instinct, but the vast majority have a hunting ability best described as laughable – unless bred specifically for the task. My dog will chase my cat across the garden at top speed, but once he corners her, the fun's over, and he wants her to run again. For her part, she just wishes he was dead.

**Winner: CATS**

## Number of toes

What do you mean I'm scraping the barrel? Toes are important. Polydactylism is relatively common in cats, but rare in dogs.

**Winner: CATS**

## Evolution

A 2015 study published in *Proceedings of the National Academy of Sciences* revealed that members of the cat family have in the past been better at surviving than members of the dog family. Dogs originated in North America about 40 million years ago and by 20 million years ago the continent was home to over 30 species. It could have been so many more, but for cats. The researchers found that **cats played a significant role in making 40 dog species extinct by outcompeting them for the available food**, whereas there was no evidence that dogs wiped out a single cat species. Different hunting methods may have been to blame for dogs' failure, as well as cats' claws, which are retractable and so remain constantly sharp. By contrast, dogs' claws don't retract and are usually blunter. Whatever the cause, the report stated that 'Felids must have been more efficient predators', which means at some level they were clearly *better.*

**Winner: CATS**

## 9.04 Why do cats hate dogs?

Both cats and dogs are recently domesticated carnivorous hunting predators, highly partial to a hunk of fresh meat, and largely retaining their urge to hunt and kill. Dogs lived alongside humans for many more years before cats came along, then all of a sudden, 10,000 years ago, they were forced to share their homes and supplies of food and affection with these stroppy little beasts. Add to that the fact that most dogs are bigger than cats, and surely you have your answer: big hungry dog want eat small crunchy cat.

But of course it's not so simple. Many homes, including mine, contain both a cat and a dog and it isn't like that at all. My cat clearly hates the dog, but he, for his part, is very fond of her, seeking her attention and wanting her to play. For her part, my cat would rather stick pins in her eyes than play with no stoopid hound dawg and seems very much in charge of the relationship. Although the dog occasionally chases the cat, it's more often the other way around, with him the submissive party kept in check with a few paw whacks and lunges. This seems to be common: a study published in the *Journal of Veterinary Behavior* found that **57% of cats were aggressive to dogs in the same home (but only 10% ever harmed the dog), and only 18% of dogs threatened cats (with only 1% ever harming the cats)**.

I can't believe my soppy dog would ever hurt my vicious cat, but then he is around eight times her size, so dual-species owners can't entirely relax. It's possible that an older dog unsocialized with cats and still in touch with its predatory roots would find a young kitten delicious and tender, so carefully guiding both animals through socialization is important to avoid unseemly snacking. Making sure kittens and puppies spend time with the other species – as well as with humans – in a safe and controlled situation during their first four to eight weeks (for kittens) and first five to 12 weeks (for puppies) is key to foster interaction and mutual trust.

## 9.05 Cat people vs dog people ...

Or 'How to upset a large proportion of the world's population in 300 words'. Look, I know that there's a great variation in personality types, so I'm not saying that all dog owners like me are definitely aggressive, overbearing, delusional egomaniacs – I'm just saying that we probably are. Hang on – that's not coming out right. Look, I'm a dog lover, cat lover, gerbil lover and human lover so I'm not biased – it's just that a 2010 University of Texas study of self-identified dog lovers and cat lovers found that **cat folk tended to be less cooperative, conscientious, compassionate and outgoing than dog people and more likely to suffer from anxiety and depression**. But while cat people were more neurotic, they were also more open, artistic and intellectually curious than dog people. In 2015 researchers in Australia found that dog owners scored higher than cat owners on traits relating to competitiveness and social dominance, which matched their predictions (because dogs are more easily dominated, they assumed that their owners tended to be people who were more dominant). But they also found that cat owners scored just as highly as dog owners on narcissism and interpersonal dominance.

In 2016 Facebook published research on its own data (so, bear in mind it's specific to Facebook users, although the company does have a creepy ability to know a lot of things about a lot of people) and found that:

- Cat people are more likely to be single (30%) than dog people (24%).

- Dog people have more friends (well, they have more Facebook connections).

- Cat people are more likely to be invited to events.

Facebook also found that cat people are more literary in the books they mention (such as *Dracula*, *Watchmen*, *Alice in Wonderland*) and dog people more dog-obsessed and religious (*Marley and Me*, *Lessons from Rocky* – both about dogs – and *The Purpose Driven Life* and *The Shack* – both about God). Dog people like soppy movies about love and sex (*The Notebook*, *Dear John*, *Fifty Shades of Grey*), whereas cat people like death, hopelessness and drugs – with a bit of love and sex thrown in (*Terminator 2*, *Scott Pilgrim vs the World*, *Trainspotting*).

But Facebook's data gets really interesting (and scarily intrusive) when it comes to emotions. It truly seems to mirror the stereotypes of their animals, finding that cat people are much more likely to express tiredness, amusement and annoyance in their online posts than dog people, whereas dog folk are more likely to express excitement, pride and 'blessedness'.

# Chapter 10: Cat Food & Drink

## 10.01 What's in cat food?

In 2020 the global pet food market was worth £54.8 billion ($74.6 billion), while the UK market alone was worth £2.9 billion ($3.7 billion). **The first commercial pet food was launched in England in the 1860s by James Spratt**, an American entrepreneur who had travelled to London to sell lightning conductors but got distracted, so the story goes, when he was given some inedible ship's biscuits for his dog. He spotted a gap in the market and came up with his Meat Fibrine Dog Cakes. Yummy. He was hugely successful, first in the UK and later in the USA, and one of his early UK employees was Charles Cruft, who eventually left to set up Crufts dog show.

Despite the horror stories, pet food companies don't just throw anything into cat food. It's a highly regulated industry and some standards are surprisingly high: the animals used as ingredients have to pass vet inspections to ensure they are fit for human consumption at time of slaughter. Pets, road kill, wild animals, lab animals and fur-bearing animals aren't allowed, nor is meat from sick or diseased animals. Cat food is usually a mixture of cow, chicken, lamb and fish off-cuts and derivatives, and by-products from food made for human consumption. This usually includes liver, kidneys, udders, tripe, trotters and lungs, which might not sound particularly appetizing to you, but cats love them. Importantly, it also means that nothing usable from a slaughtered animal goes to waste.

Although commercial cat food is mostly meat, nutritional additives such as extra taurine (an amino acid that cats can't make themselves), vitamins A, D, E, K and various minerals are also added.

Wet cat food is generally cooked into a meatloaf before being cut into chunks and mixed with either jelly or gravy. It's then put into cans, trays or pouches that are re-cooked in a retort (a huge pressure cooker) at 116–130°C (241–266°F) to kill bacteria, leaving the sealed pack surprisingly sterile and with a long shelf life. The cans are then cooled before labelling.

Dry cat food (or kibble) is more interesting. As with wet food, it starts with a mixture of meat and meat derivatives but these are usually cooked and ground into dry powders before being mixed with cereals, vegetables and nutritional additives. Water and steam are added to make a hot, thick dough and this is pushed through an extruder – a huge screw thread that compresses and heats the dough – before it's forced through a small nozzle called a die (cheese puffs are made in much the same way) and chopped into shapes by revolving blades as it squirts out. This heating degrades some of the nutrients in the meat so more of these need to be added again later. The change of pressure as the cooked dough comes out makes it puff up into kibbles and these are heated to dry out before being sprayed with flavourings and nutritional additives to replace those degraded by the whole process.

**Is cat food fit for human consumption? Well, by law, all ingredients used for pet food in the UK have to be fit for human consumption – and because they are all thoroughly cooked in a retort, they are highly unlikely to contain harmful bacteria.**

## 10.02 Why are cats so picky about their food?

Cats' food fickleness is legendary: **it's not uncommon for cats who've eaten the same food for their entire lives to suddenly refuse to eat it ever again**. One possible reason for this is food-aversion learning: if a cat falls sick, it may make a negative association between its illness and the last meal it had, whether or not the food was the cause of the sickness. This aversion can be a useful survival mechanism, and is also often irreversible – usually the owner's only choice is to change the flavour or brand of food. Another explanation could be that cats have a 'food variety mechanism': a genetic predisposition to occasionally change their diet to avoid becoming dependent on a food source that may at some point disappear.

More often, cats will for no apparent reason decide to stop eating at their regular mealtimes or turn their noses up at the delicious dish you put in front of them. This could have many causes. For cats, feeding is a time of potential vulnerability and anxiety and they're likely to be disturbed by anything unusual or uncontrollable when they eat. Factors such as another pet nearby, another cat in the garden, or a new disinfectant odour near its bowl can all make a cat feel anxious, yet its owner may be totally unaware.

If you're worried about your cat's diet, remember she's best tuned to eating small amounts (think mice) often, and if feeding her like this doesn't fit in with your schedule, it can appear she's just picking at her food. She may also be self-regulating her protein-to-fat intake (see p88) to stay in balance after snacking on something while out roaming. But there's often little to worry about unless the loss of appetite continues for more than two or three days, at which point call the vet.

## 10.03 Can a cat be vegan?

**Cats, like lions and tigers, are obligate carnivores: dedicated meat eaters that get all their nutritional requirements from meat**. There are different levels of carnivorousness in nature, from hypocarnivores, whose diet is made up of less than 30% meat, to mesocarnivores at 30–70% meat, and hypercarnivores with a diet of more than 70% meat. As you can guess, cats fit into this last category. But it doesn't mean cats can't eat plants – they might eat them if there's absolutely nothing else available. They can even get some nutrition from them, but they can't break them down properly to get all the nutrients they need to thrive*. Unlike other animals, cats need a diet containing specific nutrients found in abundance in meat, such as taurine, vitamin A and arachidonic acid.

Cats' digestive tracts are relatively short compared with those of omnivores such as humans and ruminants like sheep, simply because they don't need to be longer: meat is much easier to break down than vegetation, so why would a lean hunter waste energy on a complex metabolic-energy-intensive digestive system it doesn't need?

**It is possible to create a vegan cat diet, but it's tricky** – you need high-protein, low-fibre plant sources, flavouring that your cat will enjoy and it needs to be supplemented with taurine, thiamine, niacin, several B vitamins and many other micronutrients. Nonetheless various pet food companies now produce exactly this, though the general advice is that you should consult your vet before switching your cat to a vegan diet, and many vets are likely to disapprove because they see the traditional meat-based diet as reliable and controllable. In response, the plant-based pet food producers have claimed that those disapproving vets are 'suffering with an unshakeable ideology'. It's true that many academic studies point to adverse health effects caused by commercial meat-based cat and dog food diets, but there is no clear research proving that vegan or vegetarian food is any better.

## Fat Cat

**The heaviest cat ever recorded by *Guinness World Records* was Himmy, who weighed 21.3kg (46lb 15½oz) when he died of respiratory failure in Australia in 1986. Himmy had to be transported by wheelbarrow. There may have been bigger cats since, but Guinness stopped recording the category to deter people from over-feeding their pets just to get in the record books.**

*After all, a human can live off nothing but jam sandwiches for many years but vitamin and mineral deficiencies will eventually cause a cascade of nutritional, developmental, immune system and cardiovascular problems and the high possibility of an early death.

## 10.04 Why do cats eat grass?

Cleaning up cat vom is unpleasant but grassy vom is the worst (see p44). It's watery, it soaks with rude enthusiasm into the carpet, and it's made mainly of bubbly acrid gastric juices.

Cats, like humans, lack the digestive machinery and chemistry to metabolize cellulose-based grass, and that's why they vomit it out. But why do they eat it in the first place? One theory is that cats extract folic acid from the moisture in grass, and this is converted to vitamin B9 (also known as folate) by metabolism. That B9 is essential to the production of haemoglobin and without it cats can become anaemic. But if that's important to a cat, why would they regurgitate it before breaking it down into its component parts? Of course, it is always possible they digest a fair proportion of the grass before vomiting a small amount back up.

**Cats occasionally need to make themselves throw up** – if they're ill or need to clear indigestible matter (fur, feathers, intestinal parasites) from their digestive tract. So it's possible they use grass as an emetic – a food or medicine ingested to cause vomiting (it's used in cases of poisoning to clear out any toxins). Another theory is that grass has a laxative effect, helping the cat to have regular bowel movements. Obviously these two theories are polar opposites. One thing we know is that cats don't seem to suffer unduly from the odd vomit and that eating grass is perfectly safe for them. Better out than in.

## 10.05 Why do cats love eating fish?

Cats' legendary love of fish seems obvious at first: the high pungency attracts their sense of smell and the high protein content makes it a great macronutrient. My cat Cheeky has managed to kill almost all of our family goldfish, although she didn't eat them – just liberated them from their bowl. But cats' obsession with fish is very strange for several reasons.

First, because most cats avoid water, which is where you'd reasonably expect to find fish, so in evolutionary terms they are unlikely to have met often. Second, because fresh fish is not particularly suited to be a major part of their diet. Cats can't cope well with fish bones, and canned tuna can be high in mercury and phosphorus, which is bad for cats with kidney disease. Fish is also responsible for a large proportion of allergies in cats (around a quarter of all food allergies in one study).

Yet despite all these drawbacks, **some cats love fish so much that if they are fed it too often, they will refuse to eat anything else**. Small treats of fish are unlikely to cause problems, but do be careful or you could find yourself in an expensive culinary cul-de-sac.

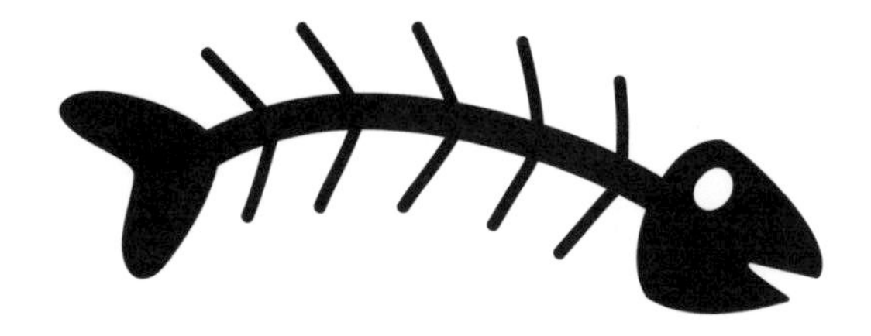

## 10.06 Why do cats stroke the floor near their food?

My badly named, badly behaved tabby Cheeky occasionally performs a strange floor-stroking action near her food bowl after finishing her supper. It looks as though she's either clearing up non-existent spilt food or shifting some invisible dirt from one place to another, but in reality she's achieving nothing – much like a teenager cleaning the dining table by stroking it with a cloth and a big dollop of resentment. The stroking can last for around two minutes until she (the cat, not the teenager) comes to her senses and wanders off.

Although this activity has no practical use, it's pretty common, and could be related to cats' impeccable hygiene. They use the same action to bury faeces or urine, so **it may be related to cats' inclination to hide evidence of their presence to avoid possible predators**. In the excellent *The Domestic Cat: The Biology of its Behaviour*, authors Dennis Turner and Patrick Bateson are at a loss to explain exactly why

cats do this, but suggest it's an evolutionary leftover: 'Such robust and evolutionary ancient forms of behaviour are not subject to the usual rules of learning whereby unrewarded activities disappear from the animal's repertoire.' In short, it's a nonessential evolutionary echo that simply hasn't faded yet.

## The Longest Cat

**Stewie the Maine Coon from Reno, USA, was vast, measuring 1m 23cm (48.5in) from the tip of his nose to the end of his tail. He was a certified therapy animal and regular visitor to his local senior home until he died in January 2013 at the age of eight.**

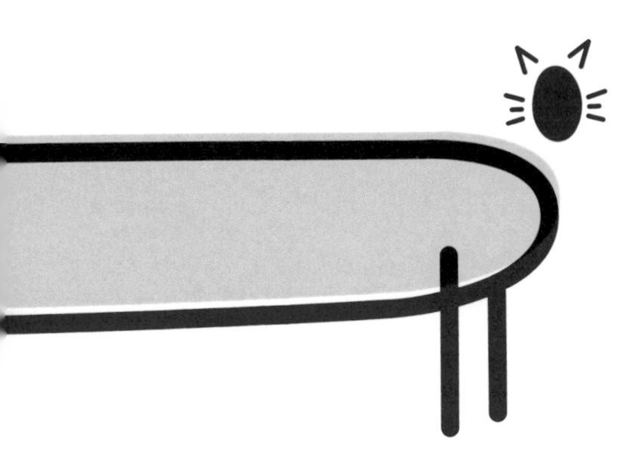

# Sources

I read an ocean of books, articles and research papers whilst writing *Catology*, and I am indebted to the wonderful authors of all of them (apologies that only a tiny fraction are listed here) despite a wide range of results, some of them downright contradictory. But that's the nature of scientific research – as the methodology changes, so does the nature of the results, and science communicators like me have to read as widely as possible, assess the relevance and context, and tread a path through the information, hoping that we aren't straying from the truth. I have tried my damndest to be very clear whether I'm reporting either scientific research or opinion, even if that opinion comes from veterinary professionals. There's so much more to learn about cats, and every new piece of research helps us understand them more and care for them better.

### General

'Pet Population 2020' (PFMA)
pfma.org.uk/pet-population-2020

### 1.01 A very unscientific introduction

'Facts + statistics: Pet statistics' (Insurance Information Institute)
iii.org/fact-statistic/facts-statistics-pet-statistics

'US pet ownership statistics' (AVMA)
avma.org/resources-tools/reports-statistics/us-pet-ownership-statistics

### 2.01 A brief history of the cat

'Phylogeny and evolution of cats (Felidae)' by Lars Werdelin, Nobuyuki Yamaguchi & WE Johnson in *Biology and Conservation of Wild Felids* by DW Macdonald & AJ Loveridge (Eds) (Oxford University Press, 2010), pp59–82
researchgate.net/publication/266755142_Phylogeny_and_evolution_of_cats_Felidae

'The near eastern origin of cat domestication' by Carlos A Driscoll *et al*, *Science* 317(5837) (2007), pp519–523
science.sciencemag.org/content/317/5837/519

### 2.02 Is your cat basically a cute tiger?

'Personality structure in the domestic cat (*Felis silvestris catus*), Scottish wildcat (*Felis silvestris grampia*), clouded leopard (*Neofelis nebulosa*), snow leopard (*Panthera uncia*), and

African lion (*Panthera leo*): A comparative study' by Marieke Cassia Gartner, David M Powell & Alexander Weiss, *Journal of Comparative Psychology* 128(4) (2014), pp414–426
psycnet.apa.org/record/2014-33195-001

**3.04 Can cats be left- or right-pawed?**

'Lateralization of spontaneous behaviours in the domestic cat, *Felis silvestris*' by Louise J McDowell, Deborah L Wells & Peter G Hepper, *Animal Behaviour* 135 (2018), pp37–43
sciencedirect.com/science/article/abs/pii/S0003347217303640#

'Laterality in animals' by Lesley J Rogers, *International Journal of Comparative Psychology* 3:1 (1989), pp5-25
escholarship.org/uc/item/9h15z1vr

'Motor and sensory laterality in thoroughbred horses' by PD McGreevy & LJ Rogers, *Applied Animal Behaviour Science* 92:4 (2005), pp337–352
sciencedirect.com/science/article/abs/pii/S0168159104002916?via%3Dihub

**3.05 The science of paws and claws**

'Feline locomotive behaviour'
veteriankey.com/feline-locomotive-behavior

'Locomotion in the cat: basic programmes of movement' by S Miller, J Van Der Burg, F Van Der Meché, *Brain Research* 91(2) (1975), pp239–53
ncbi.nlm.nih.gov/pubmed/1080684

'Biased polyphenism in polydactylous cats carrying a single point mutation: The Hemingway model for digit novelty' by Axel Lange, Hans L Nemeschkal & Gerd B Müller, *Evolutionary Biology* 41(2) (2013), pp262–75

'The Hemingway Home and Museum'
hemingwayhome.com/cats

**3.07 Why do cats have those evil-looking eyes?**

'Why do animal eyes have pupils of different shapes?' by William W Sprague, Jürgen Schmoll, Jared AQ Parnell & Gordon D Love, *Science Advances* 1:7 (2015), e1500391
advances.sciencemag.org/content/1/7/e1500391

**3.08 Why do cats always land on their feet?**

'Feline locomotive behaviour'
veteriankey.com/feline-locomotive-behavior

### 3.09 How many hairs are there on your cat?

'Cleanliness is next to godliness: mechanisms for staying clean' by Guillermo J Amador & David L Hu, *Journal of Experimental Biology* 218 (2015), 3164–3174
jeb.biologists.org/content/218/20/3164

'Weight to body surface area conversion for cats' by Susan E Fielder, *MSD Manual Veterinary Manual* (2015)
msdvetmanual.com/special-subjects/reference-guides/weight-to-body-surface-area-conversion-for-cats

### 3.12 How old is your cat?

'Feline life stage guidelines' by Amy Hoyumpa Vogt, Ilona Rodan & Marcus Brown, Journal of Feline Medicine and Surgery 12:1 (2010)
journals.sagepub.com/doi/10.1016/j.jfms.2009.12.006

### 4.01 Why does cat poo smell so bad?

'The chemical basis of species, sex, and individual recognition using feces in the domestic cat' by Masao Miyazaki *et al*, *Journal of Chemical Ecology* 44 (2018), pp364–373
link.springer.com/article/10.1007/s10886-018-0951-3

'The fecal microbiota in the domestic cat (*Felis catus*) is influenced by interactions between age and diet; a five year longitudinal study' by Emma N Bermingham, *Frontiers in Microbiology* 9:1231 (2018)
frontiersin.org/articles/10.3389/fmicb.2018.01231/full

'Gut microbiota of humans, dogs and cats: current knowledge and future opportunities and challenges' by Ping Deng & Kelly S Swanson, *British Journal of Nutrition* 113: S1 (2015), ppS6–S17
cambridge.org/core/journals/british-journal-of-nutrition/article/gut-microbiota-of-humans-dogs-and-cats-current-knowledge-and-future-opportunities-and-challenges/D0EA4D0E254DD5846613CB338295D2D3/core-reader

'About your companion's microbiome'
animalbiome.com/about-your-companions-microbiome

### 4.02 Why don't cats fart (but dogs do)?

*Fartology: The Extraordinary Science Behind the Humble Fart* by Stefan Gates (Quadrille, 2018)
gastronauttv.com/books

'The chemical basis of species, sex, and individual recognition using feces in the domestic cat' by Masao Miyazaki *et al*, *Journal of Chemical Ecology* 44 (2018), pp364–373
link.springer.com/article/10.1007/s10886-018-0951-3

**4.04 Hairball!**

'Cats use hollow papillae to wick saliva into fur' by Alexis C Noel & David L Hu, *Proceedings of the National Academy of Sciences of the United States of America* 115(49) (2018), 12377-12382

**5.02 Does your cat love you?**

'Sociality in cats: a comparative review' by John WS Bradshaw, *Journal of Veterinary Behavior* 11 (2016), pp113–124
sciencedirect.com/science/article/abs/pii/S1558787815001549?via%3Dihub

'Attachment bonds between domestic cats and humans' by Kristyn R Vitale, Alexandra C Behnke & Monique AR Udell, *Current Biology* 29:18 (2019), ppR864–R865
cell.com/current-biology/fulltext/S0960-9822(19)31086-3

'Domestic cats (*Felis silvestris catus*) do not show signs of secure attachment to their owners' by Alice Potter & Daniel Simon Mills, *PLOS ONE* 10(9) (2015), e0135109
journals.plos.org/plosone/article?id=10.1371/journal.pone.0135109

'Social interaction, food, scent or toys? A formal assessment of domestic pet and shelter cat (*Felis silvestris catus*) preferences' by Kristyn R Vitale Shreve, Lindsay R Mehrkamb & Monique AR Udell, *Behavioural Processes* 141:3 (2017), pp322–328
sciencedirect.com/science/article/abs/pii/S0376635716303424

**5.04 Are cats capable of abstract thought?**

'There's no ball without noise: cats' prediction of an object from noise' by Saho Takagi *et al*, *Animal Cognition* 19 (2016), pp1043–1047
link.springer.com/article/10.1007/s10071-016-1001-6

**5.05 Do cats dream?**

'Behavioural and EEG effects of paradoxical sleep deprivation in the cat' by M Jouvet, *Proceedings of the XXIII International Congress of Physiological Sciences* (*Excerpta Medica International Congress Series* No.87, 1965)
sommeil.univ-lyon1.fr/articles/jouvet/picps_65/

**5.07 Does your cat know when you're unhappy?**
'Empathic-like responding by domestic dogs (*Canis familiaris*) to distress in humans: an exploratory study' by Deborah Custance & Jennifer Mayer, *Animal Cognition* 15 (2012), pp851–859
ncbi.nlm.nih.gov/pubmed/22644113?dopt=Abstract

'Man's other best friend: domestic cats (*F. silvestris catus*) and their discrimination of human emotion cues' by Moriah Galvan & Jennifer Vonk, *Animal Cognition* 19 (2015), pp193–205
link.springer.com/article/10.1007/s10071-015-0927-4

**5.08 Where does your cat go once he's out of the catflap?**
'Roaming habits of pet cats on the suburban fringe in Perth, Western Australia: What size buffer zone is needed to protect wildlife in reserves?' by Maggie Lilith, MC Calver & MJ Garkaklis, *Australian Zoologist* 34 (2008), pp65–72
researchgate.net/publication/43980337_Roaming_habits_of_pet_cats_on_the_suburban_fringe_in_Perth_Western_Australia_What_size_buffer_zone_is_needed_to_protect_wildlife_in_reserves

**5.09 What does your cat do at night?**
'The use of animal-borne cameras to video-track the behaviour of domestic cats' by Maren Huck & Samantha Watson, *Applied Animal Behaviour Science* 217 (2019), pp63–72
sciencedirect.com/science/article/abs/pii/S0168159118306373

'Daily rhythm of total activity pattern in domestic cats (*Felis silvestris catus*) maintained in two different housing conditions' by Giuseppe Piccione *et al*, *Journal of Veterinary Behavior* 8:4 (2013), pp189–194
sciencedirect.com/science/article/abs/pii/S1558787812001220?via%3Dihub

**5.11 Are cats really able to find their way home from miles away?**
'The homing powers of the cat' by Francis H Herrick, *The Scientific Monthly* 14:6 (1922), pp525–539
jstor.org/stable/6677?seq=1#metadata_info_tab_contents

**5.12 Why are cats scared of cucumbers?**
'Object permanence in cats and dogs' by Estrella Triana & Robert Pasnak, *Animal Learning & Behavior* 9 (1981), pp135–139
link.springer.com/article/10.3758%2FBF03212035

**5.16 Why do cats love sitting in boxes?**
'Will a hiding box provide stress reduction for shelter cats?' by CM Vinkea, LM Godijn & WJR van der Leij, *Applied Animal Behaviour Science* 160 (2014), pp86–93
sciencedirect.com/science/article/abs/pii/S0168159114002366

**5.17 Why do cats make great mums but terrible dads?**
'Aggression in cats'
aspca.org/pet-care/cat-care/common-cat-behavior-issues/aggression-cats

**6.01 How do cats see in the dark?**
'Electrophysiology meets ecology: Investigating how vision is tuned to the life style of an animal using electroretinography' by Annette Stowasser, Sarah Mohr, Elke Buschbeck & Ilya Vilinsky, *Journal of Undergraduate Neuroscience Education* 13(3) (2015), A234–A243
ncbi.nlm.nih.gov/pmc/articles/PMC4521742/

**6.03 How good is your cat's sense of taste?**
'Balancing macronutrient intake in a mammalian carnivore: disentangling the influences of flavour and nutrition' by Adrian K Hewson-Hughes, Alison Colyer, Stephen J Simpson & David Raubenheimer, *Royal Society Open Science* 3:6 (2016)
royalsocietypublishing.org/doi/full/10.1098/rsos.160081#d14640073e1

'Pseudogenization of a sweet-receptor gene accounts for cats' indifference toward sugar' by Xia Li *et al*, *PLOS Genetics* 1(1): e3 (2005)
journals.plos.org/plosgenetics/article?id=10.1371/journal.pgen.0010003

'Taste preferences and diet palatability in cats' by Ahmet Yavuz Pekel, Serkan Barış Mülazımoğlu & Nüket Acar, *Journal of Applied Animal Research* 48:1 (2020), pp281–292
tandfonline.com/doi/pdf/10.1080/09712119.2020.1786391

**7.01 Why do cats meow?**
'Vocalizing in the house-cat; a phonetic and functional study' by Mildred Moelk, *The American Journal of Psychology* 57:2 (1944), pp184–205
jstor.org/stable/1416947?seq=1

'Domestic cats (*Felis catus*) discriminate their names from other words' by Atsuko Saito, Kazutaka Shinozuka, Yuki Ito & Toshikazu Hasegawa, *Scientific Reports* 9:5394 (2019)
nature.com/articles/s41598-019-40616-4

## 8.02 Are cats good for your health?

'To have or not to have a pet for better health?' by Leena K Koivusilta & Ansa Ojanlatva, *PLOS ONE* 1(1) (2006), e109
ncbi.nlm.nih.gov/pmc/articles/PMC1762431/

'Cat ownership and the risk of fatal cardiovascular diseases. Results from the second National Health and Nutrition Examination Study Mortality Follow-up Study' by Adnan I Qureshi, Muhammad Zeeshan Memon, Gabriela Vazquez & M Fareed K Suri, *Journal of Vascular and Interventional Neurology* 2(1) (2009), pp132–135
ncbi.nlm.nih.gov/pmc/articles/PMC3317329/

'Animal companions and one-year survival of patients after discharge from a coronary care unit' by E Friedmann, AH Katcher, JJ Lynch & SA Thomas, *Public Health Reports* 95(4) (1980), pp307–312
ncbi.nlm.nih.gov/pmc/articles/PMC1422527/

'Pet ownership and health in older adults: findings from a survey of 2,551 community-based Australians aged 60-64' by Ruth A Parslow *et al*, *Gerontology* 51(1) (2005), pp40–7
ncbi.nlm.nih.gov/pubmed/15591755

'Impact of pet ownership on elderly Australians' use of medical services: an analysis using Medicare data' by AF Jorm *et al*, *The Medical Journal of Australia* 166(7) (1997), pp376–7
ncbi.nlm.nih.gov/pubmed/9137285

'Are pets in the bedroom a problem?' by Lois E Krahn, M Diane Tovar & Bernie Miller, *Mayo Clinic Proceedings* 90:12 (2015), pp1663–1665, mayoclinicproceedings.org/article/S0025-6196(15)00674-6/abstract

'Multiple pets may decrease children's allergy risk'
https://www.niehs.nih.gov/news/newsroom/releases/2002/august27/index.cfm

## 8.03 Are cats BAD for your health?

'Toxoplasmosis rids its host of all fear'
unige.ch/communication/communiques/en/2020/quand-la-toxoplasmose-ote-tout-sentiment-de-peur/

'Cat-associated zoonoses' by Jeffrey D Kravetz & Daniel G Federman, *Archives of Internal Medicine* 162(17) (2002), pp1945-1952
jamanetwork.com/journals/jamainternalmedicine/fullarticle/213193

**8.04 How much does a cat cost to run?**

'The cost of owning a cat'
battersea.org.uk/pet-advice/cat-advice/cost-owning-cat

'The cost of owning a dog'
rover.com/blog/uk/cost-of-owning-a-dog/

**8.07 Why are people allergic to cats?**

'Dog and cat allergies: current state of diagnostic approaches and challenges' by Sanny K Chan & Donald YM Leung, *Allergy, Asthma & Immunology Research* 10(2) (2018), pp97–105
ncbi.nlm.nih.gov/pmc/articles/PMC5809771/

**8.08 Why do cats love people who hate cats?**

'Environmental impacts of food consumption by dogs and cats' by Gregory S Okin, *PLOS ONE* 12(8) (2017), e0181301
ournals.plos.org/plosone/article?id=10.1371/journal.pone.0181301

**8.12 What's the climate impact of a cat?**

'The climate mitigation gap: education and government recommendations miss the most effective individual actions' by Seth Wynes & Kimberly A Nicholas, *Environmental Research Letters* 12:7 (2017)
iopscience.iop.org/article/10.1088/1748-9326/aa7541

**8.13 Are cats bird-murdering machines?**

'The impact of free-ranging domestic cats on wildlife of the United States' by Scott R Loss, Tom Will & Peter P Marra, *Nature Communications* 4, 1396 (2013)
nature.com/articles/ncomms2380

'Are cats causing bird declines?'
rspb.org.uk/birds-and-wildlife/advice/gardening-for-wildlife/animal-deterrents/cats-and-garden-birds/are-cats-causing-bird-declines/

### 9.02 Cats vs Dogs: social and medical

'Pet Population 2020' (PFMA)
pfma.org.uk/pet-population-2020

'Dogs have the most neurons, though not the largest brain: trade-off between body mass and number of neurons in the cerebral cortex of large carnivoran species' by Débora Jardim-Messeder *et al*, *Frontiers in Neuroanatomy* 11:118 (2017)
frontiersin.org/articles/10.3389/fnana.2017.00118/full

### 9.03 Cats vs Dogs: physical head-to-head

'The role of clade competition in the diversification of North American canids' by Daniele Silvestro, Alexandre Antonelli, Nicolas Salamin & Tiago B Quental, *Proceedings of the National Academy of Sciences of the United States of America* 112(28) (2015), 8684-8689
pnas.org/content/112/28/8684

### 9.05 Cat people vs dog people

'Personalities of self-identified "dog people" and "cat people"' by Samuel D Gosling, Carson J Sandy & Jeff Potter, *Anthrozoös* 23(3) (2010), pp213–222
researchgate.net/publication/233630429_Personalities_of_Self-Identified_Dog_People_and_Cat_People

'Cat people, dog people' (Facebook Research)
research.fb.com/blog/2016/08/cat-people-dog-people/

### 10.01 What's in cat food?

'Identification of meat species in pet foods using a real-time polymerase chain reaction (PCR) assay' by Tara A Okumaa & Rosalee S Hellberg, *Food Control* 50 (2015), pp9–17
sciencedirect.com/science/article/abs/pii/S0956713514004666

'Pet food' (Food Standards Agency)
food.gov.uk/business-guidance/pet-food

'The history of the pet food industry'
web.archive.org/web/20090524005409/petfoodinstitute.org/petfoodhistory.htm

**10.02 Why are cats so picky about their food?**
'Balancing macronutrient intake in a mammalian carnivore: disentangling the influences of flavour and nutrition' by Adrian K Hewson-Hughes, Alison Colyer, Stephen J Simpson & David Raubenheimer, *Royal Society Open Science* 3:6 (2016)
royalsocietypublishing.org/doi/full/10.1098/rsos.160081#d14640073e1

**10.03 Can a cat be vegan?**
'Differences between cats and dogs: a nutritional view' by Veronique Legrand-Defretin, *Proceedings of the Nutrition Society* 53:1 (2007)
cambridge.org/core/journals/proceedings-of-the-nutrition-society/article/differences-between-cats-and-dogs-a-nutritional-view/A01A77BABD1B6DDD500145D7A02D67A5

# Acknowledgements

This book is based on the work of thousands of wonderful researchers and authors who have put their expertise down in print, and although I've cited the main papers and books I've referenced, there were hundreds more that were vital for understanding this fascinating world. It's both odd and sad that most of this is publicly-funded research, yet scientific publishers make huge profits from it and effectively ring-fence that knowledge from the public. Let's hope this changes sooner rather than later.

Thanks so much to the wonderful Sarah Lavelle, Stacey Cleworth and Claire Rochford at Quadrille for such enthusiasm for my weird fascinations and for putting up with both me and my inability to take deadlines seriously. And to Luke Bird for taking on another weird project with such good grace.

Thanks so much to my gorgeous girls Daisy, Poppy and Georgia, for leaving me alone at the end of the garden to write, and for enduring the breathlessly enthusiastic stream of facts I subjected them to over dinner. Thanks also to Blue and Cheeky for enduring my constant poking whilst testing vomeronasal organs, nictitating membranes, fur-counts, cross-species communication and claw-retracability. Thanks also to Brodie Thomson, Eliza Hazlewood and Coco Ettinghausen and, as always, to the amazing and wonderfully supportive crew at DML: Jan Croxson, Borra Garson, Lou Leftwich and Megan Page.

Lastly, thanks so much to the brilliant audiences who've come along to my shows and laughed their pants off whilst we've explored some utterly fascinating or revolting science live on stage. I love you people.

# Index

Publishing Director: Sarah Lavelle
Head of Design: Claire Rochford
Designer and Illustrator: Luke Bird
Editor: Stacey Cleworth
Copy Editor: Nick Funnell
Editorial Assistant: Sofie Shearman
Head of Production: Stephen Lang
Production Controller: Katie Jarvis

First published in 2021 by Quadrille, an imprint of Hardie Grant Publishing

Quadrille
52–54 Southwark Street
London SE1 1UN
quadrille.com



Cataloguing in Publication Data: a catalogue record for this book is available from the British Library.

ISBN: 978 1 78713 632 8
Printed in China